ADVANCES IN FRESHWATER AQUACULTURE

Technology for Sustainable Development

ADVANCES IN FRESHWATER AQUACULTURE

Technology for Sustainable Development

Tarang Kumar Shah, PhD
Madhu Sharma, PhD
Mohd. Ashraf Rather, PhD
Avdhesh Kumar, PhD
Editors

AAP | APPLE ACADEMIC PRESS

First edition published 2027

Apple Academic Press Inc.
1265 Goldenrod Circle, NE,
Palm Bay, FL 32905 USA

760 Laurentian Drive, Unit 19,
Burlington, ON L7N 0A4, Canada

CRC Press
2385 NW Executive Center Drive,
Suite 320, Boca Raton FL 33431

4 Park Square, Milton Park,
Abingdon, Oxon, OX14 4RN, UK

Library and Archives Canada Cataloguing in Publication

CIP data on file with Canada Library and Archives

Library of Congress Cataloging-in-Publication Data

CIP data on file with US Library of Congress

ISBN: 978-1-77964-084-0 (hbk)
ISBN: 978-1-77964-085-7 (ebk)
DOI: 10.1201/9781779640857

About the Editors

Tarang Kumar Shah, PhD
Assistant Professor, Department of Fisheries, DGCN College of Veterinary and Animal Sciences, Chaudhary Sarwan Kumar, Himachal Pradesh Krishi Vishvavidyalaya, Palampur, Himachal Pradesh, India

Tarang Kumar Shah, PhD, works as an Assistant Professor in the Department of Fisheries of CSK Himachal Pradesh Agricultural University, Palampur, Himachal Pradesh, India. His areas of research interest are aquaculture production and fish disease. He has published more than 20 research papers and review papers, six books, five book chapters, and many popular articles and pamphlets. Dr. Shah has been honored with several awards, including a university gold medal, Bhamashah Award, UGC University Rank Holder Fellowship, university-level fellowship, best poster presentation, young scientist award, best thesis award, etc. He completed his BFSc and MFSc (Aquaculture) at Maharana Pratap University of Agriculture and Technology, Udaipur, and his PhD (Aquaculture) at Govind Ballabh Pant University of Agriculture and Technology, Pantnagar, India.

Madhu Sharma, PhD
Assistant Professor, Department of Fisheries, DGCN College of Veterinary and Animal Sciences, Chaudhary Sarwan Kumar, Himachal Pradesh Krishi Vishvavidyalaya, Palampur, Himachal Pradesh, India

Madhu Sharma, PhD, is an Assistant Professor in the Department of Fisheries of CSK Himachal Pradesh Agricultural University, Palampur, Himachal Pradesh, India. She has 12 years of experience in teaching and research at different institutes of repute. She worked for six months at the Aquatic Lab at Adelaide University, South Australia. Her areas of research are toxicology, cytogenetics, and aquaculture. She has 36 research papers, 45 extension publications, one book, and two book chapters to her credit. She has mentored four master's students. She has received many awards and scholarships as well as a DST fellowship. She received her BSc degree from Hans Raj Mahila Maha Vidyalaya Jalandhar, Punjab. She earned her master's and PhD degrees at G.N.D.U. Amritsar, Punjab, India.

Mohd. Ashraf Rather, PhD
Assistant Professor, Division of Fish Genetics and Biotechnology, Faculty of Fisheries, Rangil, Ganderbal, SKUAST-K, Jammu and Kashmir, India

Mohd. Ashraf Rather, PhD, works as an Assistant Professor/Junior Scientist in the Division of Genetics and Biotechnology at the Faculty of Fisheries, Rangil, Ganderbal, Sher-e-Kashmir University of Agricultural Sciences and Technology, Kashmir, India. He graduated from College of Fisheries, Ratnagiri, Maharashtra, and obtained his master's and doctoral degrees in Fisheries Biotechnology from the Division of Genetics and Biotechnology, Central Institute of Fisheries Education (CIFE), Mumbai. His area of research interest is reproductive physiology and molecular endocrinology of fish proteins. His major research contributions are in the identification, characterization, and expression profiling of more than 25 reproductive genes in fish including kisspeptin genes in Indian major carps. He has more than eight years of experience of working on molecular endocrinology of fish.

Dr. Rather is a recipient of various accolades, such as Prof. K. H. Alikunhi gold medal from the Central Institute of Fisheries Education and Dr. Karunasagar Best Post-Graduate Thesis (PhD—Indian category) award from the Professional Fisheries Graduates Forum (PFGF). He has received a Young Scientist Award in 2017 from the Society of Fisheries and Life Science, College of Fisheries, Mangalore. He has also received an Asia-Pacific Alltech Young Scientist Award—2012 (first rank) in Agriculture and Allied Field from Alltech, USA. He is a recipient of various national and international fellowships, such as the Prince Songkla Fellowship from Prince Songkla University of Thailand; National Post-Doc fellowship (NPDF) from the Science and Engineering Research Board (SERB), Department of Science and Technology, New Delhi; Indian Science Academy Fellowship; Senior Research Fellowship (SRF, first rank); and Junior Research Fellowship (JRF, first rank) from the Indian Council of Agricultural Research (ICAR), New Delhi.

He has published more than 35 research publications in both national and international peer-reviewed research journals. In addition, he has published one book, two conference proceedings, and two book chapters. He functions as editorial board member for various international and national research journals. To date, Dr. Rather has received three best research paper awards from various international and national institutes and organizations. He has mentored several master's and doctoral students pursuing higher studies in fisheries biotechnology/fisheries science.

Avdhesh Kumar, PhD
Dean College of Fisheries, GB Pant University of Agriculture and Technology, Pantnagar, Uttrakhand, India

Avdhesh Kumar, PhD, is presently working as Dean in the College of Fisheries, GB Pant University of Agriculture and Technology, Pantnagar, Uttrakhand, India. He holds an undergraduate degree and master's degree from the University of Agra University, Agra. He also completed his PhD in the Zoology from Dr. B.R. Ambedkar (Agra) University. His areas of specialization include taxonomy, systematics, morphology, biodiversity, ecology, disease, and parasitology. He has published more than 30 research publications in both national and international peer-reviewed research journals. In addition, he has published many popular articles and pamphlets. He has received many awards, scholarships, certificates of merit, and fellowships, etc. He has mentored several master's and doctoral students pursuing higher studies in aquaculture.

Contents

Contributors

Pinak Bamaniya
Department of Aquatic Environment Management, College of Fisheries Science, Kamdhenu University, Veraval, Gujarat, India

Ishita Bambhaniya
Department of Aquaculture, College of Fisheries Science, Kamdhenu University, Veraval, Gujarat, India

Surya Chaitanya Banuru
AAHM Division, ICAR-CIFE, Mumbai, Maharashtra, India

Mayur Bhadarka
Department of Aquatic Environment Management, College of Fisheries Science, Kamdhenu University, Veraval, Gujarat, India

Pradyut Biswas
College of Fisheries, CAU (I), Lembucherra, West Tripura, Tripura, India

Sanjay Chandravanshi
Fisheries College and Research Institute, Tamil Nadu Dr. J. Jayalalithaa Fisheries University, Thoothukudi, Tamil Nadu, India

Vivek Chauhan
Faculty of Fisheries Science, Kerala University of Fisheries and Ocean Studies, Kochi, Kerala, India

Bhagchand Chhaba
Teaching Associate, LSPN College of Fisheries, Kawardha, Chhattisgarh, India

Bhavesh Choudhary
College of Fisheries, Central Agricultural University, Lembucherra, Tripura, India

Nayan Chouhan
College of Fisheries, Central Agricultural University, Lembucherra, Tripura, India

Rajkumar Dabhi
Assistant Professor, Department of Aquaculture, College of Fisheries Science, Kamdhenu University, Navsari

Suparna Deb
College of Fisheries, CAU (I), Lembucherra, West Tripura, Tripura, India

Payel Debbarma
College of Fisheries, CAU (I), Lembucherra, West Tripura, Tripura, India

Sourabh Debbarma
Fisheries College and Research Institute, Tamil Nadu Dr. J. Jayalalithaa Fisheries University, Thoothukudi, Tamil Nadu, India

Anamika Debnath
College of Fisheries, CAU (I), Lembucherra, West Tripura, Tripura, India

Debojit Dekari
College of Fisheries, Assam Agricultural University, Raha, Assam, India

Ng Chinglembi Devi
Dr. M.G.R Fisheries College and Research Institute, Ponneri, Thiruvallur District, Tamil Nadu, India

Shiwam Dubey
Karnataka Veterinary, Animal and Fisheries Sciences University, Bidar, College of Fisheries, Mangaluru, Karnataka, India

Neha Gangan
Department of Aquatic Environment Management, Dr. Balasaheb Sawant Konkan Krishi Vidyapeeth, Dapoli, College of Fisheries, Shirgaon, Ratnagiri, Maharashtra, India

Ediga Arun Goud
College of Fisheries, Dr. Balasaheb Sawant Konkan Krishi Vidyapeeth, Ratnagiri, Maharashtra, India

Udai Ram Gurjar
RPCAU—KrishiVigyan Kendra, Parsauni, East Champaran, Bihar, India

Jhamlal
College of Fisheries, CAU (I), Lembucherra, West Tripura, Tripura, India

H. Vadher Kailash
Department of Aquaculture, College of Fisheries Science, Kamdhenu University, Veraval, Gujarat, India

Shubham Kashyap
College of Fisheries, CAU (I), Lembucherra, West Tripura, Tripura, India

Rajinder Kaur
Krishi Vigyan Kendra, Barnala, Guru Angad Dev Veterinary and Animal Sciences University, Ludhiana, Punjab, India

Binal Khalasi
Karnataka Veterinary, Animal and Fisheries Sciences University, Bidar, College of Fisheries, Mangaluru, Karnataka, India

M. Jora Kiran
Department of Aquaculture, College of Fisheries Science, Kamdhenu University, Veraval, Gujarat, India

N. Uday Kiran
Karnataka Veterinary, Animal and Fisheries Sciences University, Bidar, College of Fisheries, Mangaluru, Karnataka, India

D. Patel Krina
Department of Aquaculture, College of Fisheries Science, Kamdhenu University, Veraval, Gujarat, India

A. Vinod Kumar
P. V. Narasimha Rao Veterinary University, College of Fishery Science, Pebbair, Telangana, India

Kommu Madhu
P. V. Narasimha Rao Veterinary University, College of Fishery Science, Pebbair, Telangana, India

Sawant Apurva Nikhil
College of Fisheries, Ratnagiri, Dr. Balasaheb Sawant Konkan Krishi Vidyapeeth, Maharashtra, India

H. L. Parmar
Department of Fisheries Resource Management, College of Fisheries Science, Kamdhenu University, Veraval, Gujarat, India

Arun Bhai Patel
College of Fisheries, Central Agricultural University, Lembucherra, Tripura, India

Laxmi Prasad
Acharya Narendra Deva University of Agriculture and Technology, Kumarganj, Ayodhya, Uttar Pradesh, India

Bhatkar Varsha Rajendra
College of Fisheries, Ratnagiri, Dr. Balasaheb Sawant Konkan Krishi Vidyapeeth, Ratnagiri, Maharashtra, India

M. Dabhi Rajkumar
Department of Aquaculture, College of Fisheries Science, Kamdhenu University, Veraval, Gujarat, India

Devarshi Ranjan
College of Fisheries, Dr. Rajendra Prasad Central Agricultural University, Muzaffarpur, Bihar, India

Pramila Sahadevan
Faculty of Fisheries, Kerala University of Fisheries and Ocean Studies, Cochin, Kerala, India

Ashish Sahu
National Bureau of Fish Genetic Resources, Lucknow, Uttar Pradesh, India; Faculty of Fisheries, Kerala University of Fisheries and Ocean Studies, Cochin, Kerala, India

Goodly Samal
Dr. Rajendra Prasad Central Agricultural University, Pusa, College of Fisheries, Dholi, Bihar, India

U. K. Sarkar
National Bureau of Fish Genetic Resources, Lucknow, Uttar Pradesh, India

Jyoti Saroj
Acharya Narendra Deva University of Agriculture and Technology, Kumarganj, Ayodhya, Uttar Pradesh, India

Panchakarla Sedyaaw
College of Fisheries, Ratnagiri, Dr. Balasaheb Sawant Konkan Krishi Vidyapeeth, Ratnagiri, Maharashtra, India

Tarang Kumar Shah
Department of Fisheries, DGCN College of Veterinary and Animal Sciences, CSKHPKV, Palampur, Himachal Pradesh, India

Madhu Sharma
Department of Fisheries, DGCN College of Veterinary and Animal Sciences, CSKHPKV, Palampur, Himachal Pradesh, India

Arya Singh
College of Fisheries, Acharya Narendra Deva University of Agriculture and Technology, Ayodhya, India

Mahender Singh
National Bureau of Fish Genetic Resources, Lucknow, Uttar Pradesh, India

Guntapalli Sravani
Division of Aquaculture, ICAR-CIFE, Mumbai, Maharashtra, India

B. Sureandiran
Department of Fisheries Resource Management, College of Fisheries Science, Kamdhenu University, Veraval, Gujarat, India

Nitin Suyani
Department of Fisheries Resource Management, College of Fisheries Science, Kamdhenu University, Veraval, Gujarat, India

Suman Takar
MPUAT—College of Fisheries Udaipur, Rajasthan, India

Vivek R. Tandel
Department of Aquatic Environment Management, College of Fisheries Science, Veraval, Kamdhenu University, Gujarat, India

Ritika A. Tandel
Department of Fish Processing Technology, College of Fisheries Science, Veraval, Kamdhenu University, Gujarat, India

Bhavika R. Tandel
Department of Fisheries Resource Management, College of Fisheries Science, Veraval, Kamdhenu University, Gujarat, India

P. S. Tanwar
Krishi Vigyan Kendra, Barnala, Guru Angad Dev Veterinary and Animal Sciences University, Ludhiana, Punjab, India

D. T. Vaghela
Department of Aquatic Environment Management, College of Fisheries Science, Kamdhenu University, Veraval, Gujarat, India

Gavhane Shrinath Vaijanath
Fisheries College and Research Institute, Tamil Nadu Dr. J. Jayalalithaa Fisheries University, Thoothukudi, Tamil Nadu, India

Aitwar Vaijnath
Karnataka Veterinary, Animal and Fisheries Sciences University, Bidar, College of Fisheries, Mangaluru, Karanataka, India

S. Vijayakumar
Fisheries Research and Information Centre, Vijayapur, Karnataka, India

Suhas Wasave
College of Fisheries, Ratnagiri, Dr. Balasaheb Sawant Konkan Krishi Vidyapeeth, Maharashtra, India

Sangita Wasave
College of Fisheries, Dr. Balasaheb Sawant Konkan Krishi Vidyapeeth, Ratnagiri, Maharashtra, India

Nitesh Kumar Yadav
College of Fisheries, Central Agricultural University, Lembucherra, Tripura, India

B. Motivaras Yagnesh
Department of Fisheries Resource Management, College of Fisheries Science, Kamdhenu University, Veraval, India

Yilbong Yirang
ICAR—Central Institute of Fisheries Education, Mumbai, Maharashtra, India

Abbreviations

BMPs	best management practices
BPA	bisphenol A
CNTs	carbon nanotubes
CRISPR	clustered regularly interspaced short palindromic repeats
DO	dissolved oxygen
eDNA	environmental DNA
EIA	Environmental Impact Assessment
EOCs	emerging organic contaminants
eRNA	environmental RNA
FAO	Food and Agriculture Organization
FOS	fructooligosaccharide
FTIR	Fourier transform infrared spectroscopy
GISs	geographic information systems
GWP	Global Water Partnership
HABs	harmful algal blooms
HGT	horizontal gene transfer
HPS	hydroponic production system
IMTA	integrated multitrophic aquaculture
IWRM	integrated water resources management
MNP	micro- and nanoplastics
MPAs	marine protected areas
MPs	microplastics
NGS	next-generation sequencing
NPs	nanoplastics
PA	polyamide
PCR	polymerase chain reaction
PE	polyethylene
PET	polyethylene tetrapthalate
PFZ	potential fishing zones
PP	polypropylene
PS	polystyrene
PUFAs	polyunsaturated fatty acids
PVC	polyvinyl chloride

QS	quorum sensing
RAS	recirculating aquaculture systems
RS	remote sensing
SCFAs	short-chain fatty acids
SEM	scanning electron microscopy
SeNPs	selenium nanoparticles
SM	secondary metabolites
SNPs	single nucleotide polymorphisms
SPF	specific pathogen-free
SPR	specific pathogen resistant
SPT	specific pathogen tolerant
SWOT	strengths, weaknesses, opportunities, and threats
WSSV	white spot syndrome virus
WWTP	wastewater treatment plant

Preface

Despite the challenges posed by the global COVID-19 pandemic in 2020, the aquaculture industry continued to grow, albeit with variations in different regions and countries. With huge food demands for a steadily growing population in 2050, the advancement of aquaculture production methods is vital. The aquaculture industry is continuously incorporating modern farming techniques and biotechnology to overcome challenges related to juvenile supply from hatcheries, nutritional quality, and other production constraints.

This book, *Advances in Freshwater Aquaculture: Technology for Sustainable Development,* serves as a comprehensive and meticulously structured guide that delves into the various methodologies and innovations in the field of aquaculture. It explores the integration of both established practices and cutting-edge technologies aimed at enhancing the efficiency and sustainability of freshwater aquaculture systems. The contents of this book encompass a broad spectrum of topics, ranging from the current state of freshwater aquaculture in India and around the world, to the advancements in aquaculture technology, the role of probiotics in aquaculture, environmental DNA surveillance, the impact of climate change on aquaculture, and the effects of microplastic in aquaculture. Aquaculture faces significant hurdles in terms of nutrition, especially concerning finfish, as outlined in discussions on important nutritional matters. Foundational knowledge is crucial for farmers to address these issues effectively. Disease management is another critical aspect of biosecurity to safeguard the continuity and sustainability of aquaculture. This publication provides a complete overview of the field of modern aquaculture, featuring a diverse group of esteemed editors and expert contributors.

We would like to extend our deepest thanks to all the authors who have played a significant role in the creation of the book titled, *Advances in Freshwater Aquaculture: Technology for Sustainable Development,* in the dynamic field of aquaculture research. The latest edition of this publication is easily accessible to undergraduate, postgraduate, and doctoral students, as well as entrepreneurs and individuals seeking to enhance their academic knowledge. It presents aquaculture information in a unique and innovative way by incorporating insights from previous studies. This collaborative endeavor has resulted in a comprehensive compilation that explores the most recent advancements and promising prospects in the field of aquaculture research.

CHAPTER 1

Present Status of Freshwater Aquaculture in India and Globe: An Overview

NITESH KUMAR YADAV[1*], NG CHINGLEMBI DEVI[2], SOURABH DEBBARMA[3], ARUN BHAI PATEL[1], GAVHANE SHRINATH VAIJANATH[3], YILBONG YIRANG[4], and SANJAY CHANDRAVANSHI[3]

[1]*College of Fisheries, Central Agricultural University, Lembucherra, Tripura, India*

[2]*Dr. M.G.R Fisheries College and Research Institute, Ponneri, Thiruvallur District, Tamil Nadu, India*

[3]*Fisheries College and Research Institute, Tamil Nadu Dr. J. Jayalalithaa Fisheries University, Thoothukudi, Tamil Nadu, India*

[4]*ICAR-Central Institute of Fisheries Education, Mumbai, Maharashtra, India*

*Corresponding author

ABSTRACT

The global population has surpassed 8 billion in 2024, presenting significant challenges in ensuring nutritionally balanced food for all. As living costs rise, so does the price of food, making it increasingly difficult to meet dietary needs worldwide. Fish emerges as a promising solution for providing affordable, nutritious meals. However, traditional capture fisheries, once a major global fish source, have now plateaued. Aquaculture offers significant promise to meet global food demand. India, as the world's second-largest contributor to aquaculture production, boasts abundant fisheries resources

Advances in Freshwater Aquaculture: Technology for Sustainable Development. Tarang Kumar Shah, Madhu Sharma, Mohd. Ashraf Rather, & Avdhesh Kumar (Eds.)

DOI: 10.1201/9781779640857-1

encompassing freshwater and brackish water reserves. In recent years, inland aquaculture has taken precedence as the primary contributor to India's total fish production. This chapter aims to provide a comprehensive overview of the current status of freshwater aquaculture, focusing on developments and trends both in India and globally. By examining the advancements, challenges, and innovations in freshwater aquaculture practices, this research contributes valuable information to the understanding and development of sustainable aquaculture strategies to address global food security challenges.

1.1 INTRODUCTION

The increasing human population has a detrimental effect on the quality of life for numerous individuals, especially in developing and underdeveloped nations (Naik, 2017). Fish represent a crucial nutritional resource for humans, providing essential nutrients such as protein, vital fatty acids, trace elements, calcium, vitamin A and B_{12}, and omega-3 fatty acids (Mohan Dey et al., 2005; Chen and Gao, 2023). Their significance extends to supporting global food security (Tacon and Metian, 2013). Fish is the primary animal protein source for billions of people worldwide. Both capture fisheries and aquaculture are vital to the livelihoods of over 10% of the global population, providing essential income, employment, and food security. These sectors play a crucial role in sustaining coastal and rural communities, contributing significantly to economic stability and nutrition worldwide (Ngasotter et al., 2020; Yadav et al., 2024). Fish and seafood are considered essential components of traditional cuisine across numerous Asian and Pacific regions. Historically, the primary source of fish and seafood has been capture fishing, conducted in both oceanic and freshwater river environments, supplemented by certain forms of fish farming practiced in widely utilized ponds (Suzuki, 2021).

In recent years, aquaculture has substantially contributed to food production and meeting nutritional needs (Boyd et al., 2020). Aquaculture is experiencing swift expansion within the fisheries industry in India. Aquaculture fulfills vital dietary needs for human consumption and presents significant employment and income-generating prospects, particularly in economically disadvantaged rural regions (Jayasankar, 2018). Aquaculture is a substantial industry in India, ranking second globally after China (Annual Report, 2023). The rapid expansion in this area has been propelled by several factors such as existing aquaculture practices, population, economic growth, a flexible regulatory framework, and increasing export prospects (Bostock et al., 2010). Freshwater aquaculture pertains to the cultivation and propagation of aquatic organisms in freshwater settings, serving as a crucial component of the aquaculture sector (Wu et al., 2022). Freshwater aquaculture significantly benefits impoverished rural communities in many developing countries by enhancing food security and improving the livelihoods of those in need (FAO, 2000). India is endowed with 3.15 million ha of reservoirs, 2.45 million ha of ponds and tanks, 1.2 million ha of floodplain lakes, along with 0.28 million km of rivers and canals. Freshwater aquaculture, which held a 34% share of inland fisheries in the mid-1980s, has surged to approximately 76% more recently (Annual Report, 2023).

India's fisheries and aquaculture industry holds significant importance in food production, ensuring nutritional security in the food supply, contributing to agricultural exports, and employing approximately 14 million individuals across various roles (Mishra et al., 2017). India ranks as the third-largest fish-producing nation globally, representing 7.96% of the total global production. The estimated total fish production for the FY 2022–2023 stands at 16.25 MMT, with the inland sector contributing 12.12 MMT and the marine industry contributing 4.13 MMT. Over the past five years, the fisheries sector has maintained an annual average growth rate of 7%. The fisheries sector is of significant importance to the national economy. The share of the fisheries sector in the total gross value added (GVA), at constant prices, for 2022–2023 is estimated at Rs. 137,716 crores, constituting approximately 1.09% of the total national GVA and 6.72% of agricultural GVA (Annual Report, 2023).

1.2 STATUS OF GLOBAL AQUACULTURE

The Food and Agriculture Organization's (FAO) "The State of World Fisheries and Aquaculture, 2022" indicates that fisheries and aquaculture production hit a record high of 214 million tons in 2020. This total included 178 million tons of aquatic animals and 36 million tons of algae, reflecting a 3% increase from the 2018 figure of 213 million tons. The estimated first sale value of global aquatic animal production was US$406 billion, with aquaculture accounting for US$265 billion and capture fisheries for US$141 billion. Notably, aquaculture's contribution to global aquatic animal production reached an all-time high of 49.2% in 2020, driven by significant growth in countries such as Chile, China, and Norway. While global aquaculture production increased in all regions except Africa—where a downturn in Egypt and Nigeria was observed—the rest of Africa experienced a 14.5% growth compared with 2019. Asia continued to lead in global aquaculture, making up 91.6% of the total production (FAO, 2022).

According to the FAO in 2022, the global aquaculture production reached 122.6 million tons of live weight in 2020, reflecting an increase of 6.7 million tons since 2018. Of this total, 87.5 million tons were derived from aquatic animals, 35.1 million tons from algae, and 700 t from shells and pearls. Notably, approximately 54.4 million tons were produced in inland waters, while marine and coastal aquaculture contributed 68.1 million tons. The production of farmed finfish amounted to 57.5 million tons, with a total value of US$146.1 billion. This figure included 49.1 million tons (valued at US$109.8 billion) from inland aquaculture and 8.3 million tons (valued at US$36.2 billion) from mariculture and coastal aquaculture. Additionally, the production of other farmed aquatic species included 17.7 million tons of mollusks (US$29.8 billion), primarily bivalves; 11.2 million tons of crustaceans (US$81.5 billion); 525,000 t of aquatic invertebrates (US$2.5 billion); and 537,000 t of semiaquatic species such as turtles and frogs (US$5 billion). The growth rate of the global aquaculture production for animal species was recorded at 2.7% in 2020 compared with 2019, representing the lowest annual growth rate in over four decades (Figures 1.1 and 1.2).

1.3 STATUS OF GLOBAL INLAND AQUACULTURE/FRESHWATER AQUACULTURE

In 2020, the global inland aquaculture production reached 54.4 million tons, representing 44.4% of the world's total production of animal species and algae. Inland aquaculture employs a variety of facilities and culture methods to raise aquatic organisms. The most prevalent method for finfish culture involves the construction of earthen ponds that are widely used due to their cost-effectiveness and suitability for a range of species. Additionally, cage culture and, to a lesser extent, pen culture are also commonly practiced in inland aquaculture. However, the importance and prevalence of these methods vary significantly among different countries, influenced by local environmental conditions, economic factors, and technological advancements (FAO, 2022). These diverse cultural methods highlight the adaptability and innovation within the inland aquaculture sector, enabling it to contribute substantially to global food security and economic development.

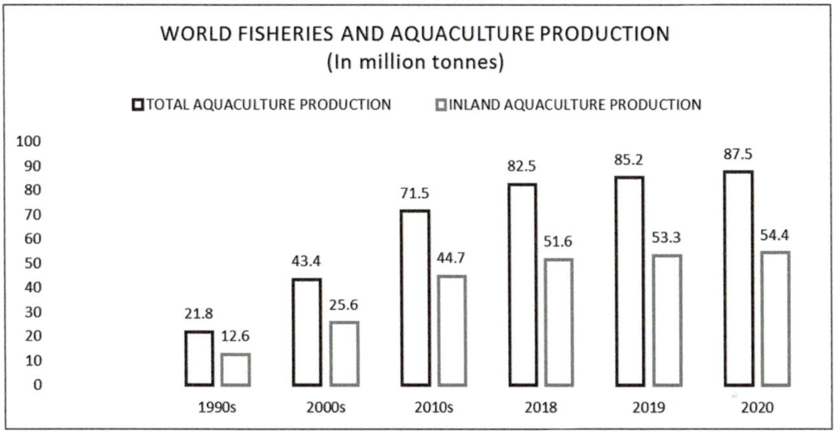

FIGURE 1.1 World fisheries and aquaculture production (in million tons).

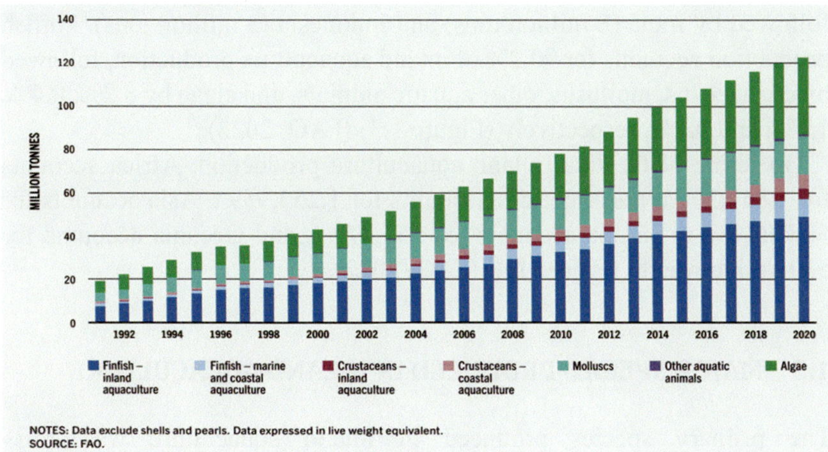

FIGURE 1.2 World Aquaculture Production (1991–2020)

Source: Reprinted from FAO (2022).

The policies at both national and local levels regarding the management of access to and the use of public open water bodies for aquaculture, particularly in relation to cage and pen culture, exhibit considerable variation among countries. With appropriate regulatory frameworks, investments in cage culture in public water bodies have been shown to be a productive and efficient means of boosting aquaculture output, thereby complementing pond culture and other aquaculture practices. In nations such as the Philippines and Indonesia, the development of cage and pen culture (including enclosures) in rivers, lakes, and reservoirs has been ongoing for several decades. However, in recent years, authorities have begun to implement measures aimed at reducing cage culture in certain water bodies to address pressing environmental and sustainability challenges (FAO, 2022).

1.4 WORLD INLAND AQUACULTURE PRODUCTION OF FINFISH BY MAJOR PRODUCERS

In 2020, global inland aquaculture production reached 54.4 million tons, representing 44.4% of the world's total aquaculture production of animal species and algae. Inland farming of aquatic animal species alone constituted 62.2% of the total aquaculture production (FAO, 2022). The world inland finfish production was topped by China with 26 million tons, followed by India (8 million tons) and Indonesia (3 million tons). Finfish production accounts for 90.2% of inland aquaculture production, followed by crustaceans, mollusks, other aquatic animals, and algae by 8.2%, 0.4%, 1.1%, and 0.1%, respectively (Figure 1.3) (FAO, 2022).

In terms of the total inland aquaculture production, Africa accounts for 1,857,361 t, the Americas accounts for 1,253,959 t, Asia accounts for 50,776,437 t, Europe accounts for 555,472 t, and Oceania accounts for 5301 t live weight (Figure 1.4) (FAO, 2022).

1.5 MAJOR SPECIES PRODUCED IN INLAND AQUACULTURE

The primary species produced in inland aquaculture was grass carp (*Ctenopharyngodon idellus*) (11.8%), followed by silver carp (*Hypophthalmichthys molitrix*) (10%) and Nile tilapia (*Oreochromis niloticus*) (9%) (Figure 1.5).

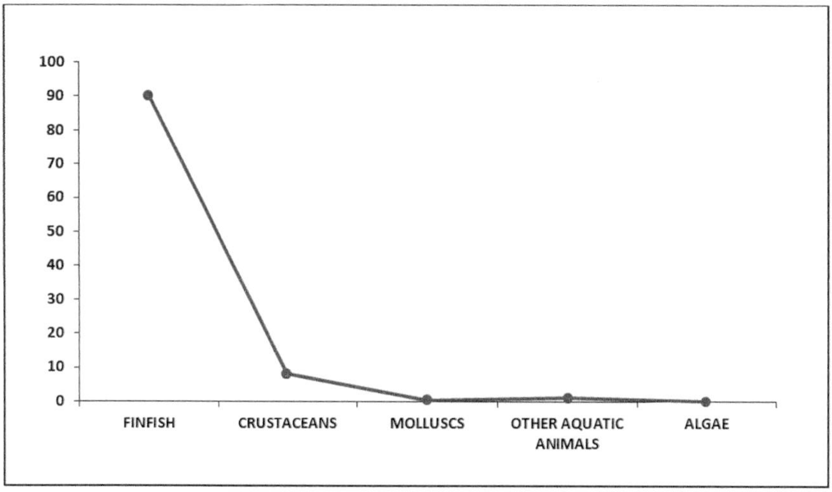

FIGURE 1.3 Inland aquaculture species production (share in %).

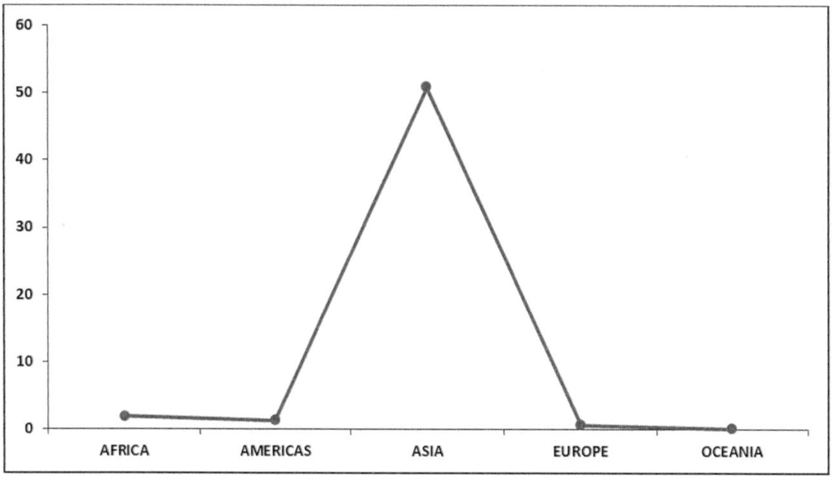

FIGURE 1.4 Inland aquaculture production in million tons by region (in lakh tons live weight).

1.6 OVERVIEW OF INDIAN AQUACULTURE

Over the years, the fisheries sector in India has undergone significant evolution, emerging as a crucial socioeconomic element for the country.

Currently, India stands as the third-largest producer of fish and aquaculture, responsible for around 16% of the total inland fish production and 5% of the global marine fish production (Department of Fisheries, GOI, 2023).

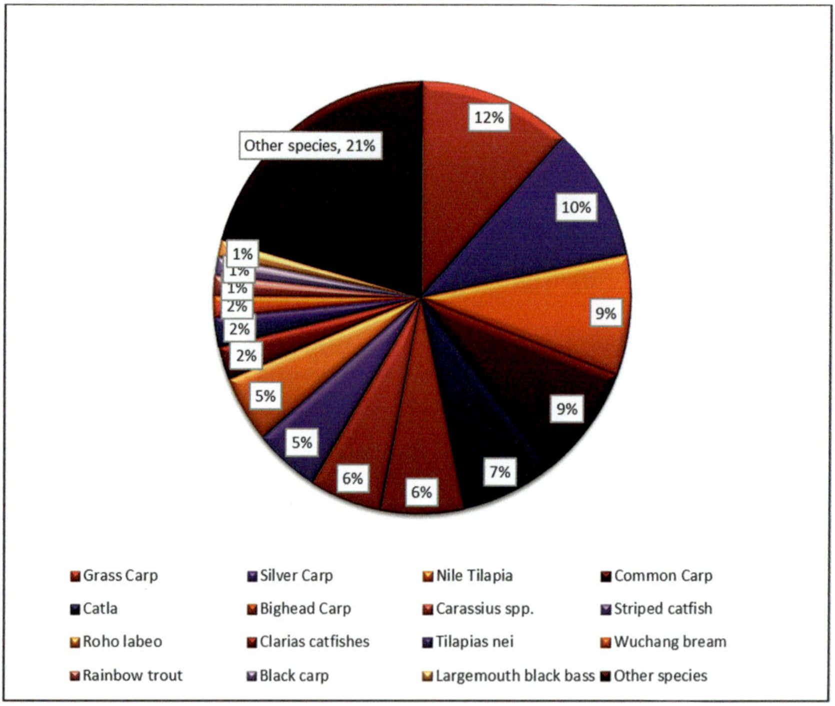

FIGURE 1.5 World major inland finfish production in 2020 (in percent).

India is among the world's top five fish exporting countries, third place preluded by China and Indonesia. About 1.1% of the contribution is to the fisheries sector in the Indian economy, and about 6.72% of the contribution is to agriculture. In 2022–2023, the country exported 1.73 MMT of seafood worth US$8.09 billion, which is an all-time high export by value. The fish production in India has surged from 24.42 lakh tons in the FY 1980–1981 to 175.45 lakh tons in 2022–2023, comprising 44.32 lakh tons marine and 131.13 lakh tons inland. The average annual growth rate of fish production stands at 7.98%. States like Andhra Pradesh, West Bengal, and Karnataka evolved into India's three major fish-producing states during 2022–2023. India's contribution to the world fish production reached 8.92% in 2022–2023 GVA by the fisheries sector increased to Rs. 2,88,526 crore in 2021–2022 from a GVA of Rs. 252,512 in 2020–2021 (Figure 1.6) (Department of Fisheries, GOI, 2023).

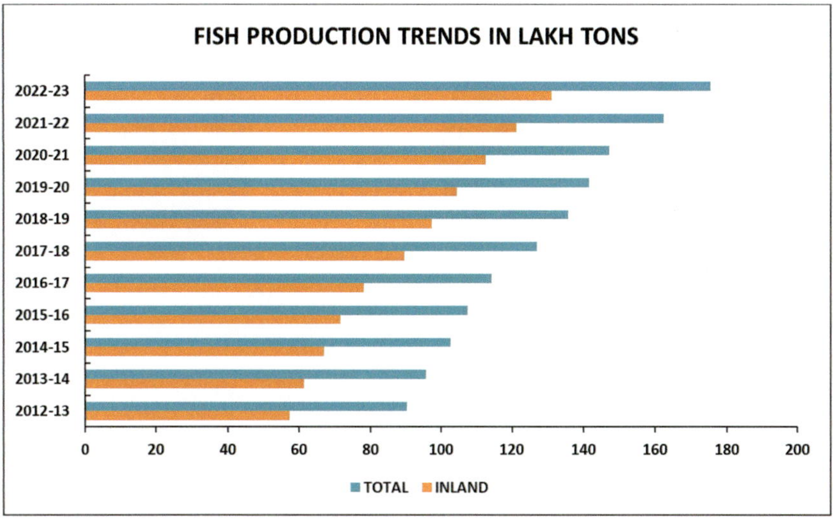

FIGURE 1.6 Fish production trends in lakh tons.

1.7 STATUS OF INLAND FISHERIES IN INDIA

Inland fish production makes up approximately 75% of the total fish production in the country, and the annual growth rate of production has also been notably high (Annual Report, 2023). India was the top producer of inland water catches in 2020, with 1.8 million tons. Of 175.45 lakh tons of the total fish production, 131.13 lakh tons come from inland production and 44.32 lakh tons from marine production. The annual fish production growth rate was estimated to be 7.98%. 8.636 million tons were from aquaculture. The annual fish production is expected to produce 3.995 million tons by 2030 (Figure 1.7) (FAO, 2022).

1.8 CONCLUSION

India possesses substantial potential in freshwater aquaculture, bolstered by its diverse array of fish species such as Indian major carps, catfishes,

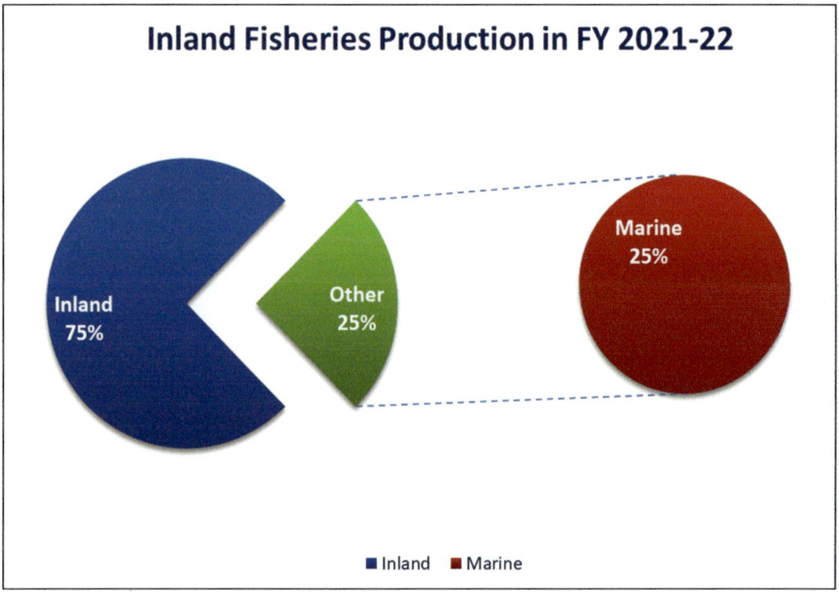

FIGURE 1.7 Inland fisheries production in FY 2021–2022.

cold water fishes, and various other commercially important species. The country's abundance of reservoirs, rivers, canals, and tanks offers extensive areas for freshwater aquaculture development. To fully capitalize on these resources, effective utilization strategies must be implemented. A comprehensive report on the current status of freshwater aquaculture will be instrumental in driving significant development and monitoring various critical issues. By understanding the existing landscape, including production levels, challenges, and opportunities, policymakers and stakeholders can make informed decisions to optimize freshwater aquaculture growth. Investments in infrastructure, technology, and training programs tailored to the specific needs of freshwater aquaculture can enhance production, promote sustainability, and contribute to the socioeconomic development of rural communities. It is essential to prioritize research and development initiatives that address key issues such as genetic improvement, disease management, water quality optimization, and market access. Furthermore, promoting best practices in aquaculture management, including sustainable farming techniques and environmental stewardship, will be pivotal in ensuring the long-term viability and success of India's freshwater aquaculture sector. By leveraging its natural resources and implementing strategic interventions, India can emerge as a global leader in sustainable freshwater aquaculture, meeting domestic demand while also contributing to international markets.

KEYWORDS

- **aquaculture**
- **production**
- **freshwater**
- **India**
- **globe**

REFERENCES

Agriculture Organization of the United Nations. (2000). Fisheries department. In: *The State of World Fisheries and Aquaculture, 2000*. Food & Agriculture Org., (Vol. 3).

Annual Report. (2023). Ministry of fisheries, animal husbandry and dairying chrome extension://efaidnbmnnnibpcajpcglclefindmkaj/https://dof.gov.in/sites/default/files/2023-04/Final_Annual_Report_2022-23_English.pdf

Bostock, J., McAndrew, B., Richards, R., Jauncey, K., Telfer, T., Lorenzen, K., Little, D., Ross, L., Handisyde, N., Gatward, I., & Corner, R. (2010). Aquaculture: Global status and trends. *Philosophical Transactions of the Royal Society B: Biological Sciences*, *365* (1554), 2897–2912.

Boyd, C. E., D'Abramo, L. R., Glencross, B. D., Huyben, D. C., Juarez, L. M., Lockwood, G. S., McNevin, A. A., Tacon, A. G., Teletchea, F., Tomasso Jr, J. R., & Tucker, C. S. (2020). Achieving sustainable aquaculture: Historical and current perspectives and future needs and challenges. *Journal of the World Aquaculture Society*, *51*(3), 578–633.

Chen, W., & Gao, S. (2023). Current status of industrialized aquaculture in China: A review. *Environmental Science and Pollution Research*, *30*(12), 32278–32287.

Department of Fisheries, Govt. of India, Handbook on Fisheries Statistics. (2023). https://dof.gov.in/sites/default/files/202308/HandbookFisheriesStatistics19012023.pdf

FAO. (2022). *The State of World Fisheries and Aquaculture 2022. Towards Blue Transformation*. FAO: Rome. https://doi.org/10.4060/cc0461en

Jayasankar, P. (2018). Present status of freshwater aquaculture in India review. *Indian Journal of Fisheries*, *65*(4), 157–165.

Mishra, S. S., Rakesh, D., Dhiman, M., Choudhary, P., Debbarma, J., Sahoo, S. N., & Mishra, C. K. (2017). Present status of fish disease management in freshwater aquaculture in India: State-of-the-art-review. *Journal of Aquaculture & Fisheries*, *1*(003), 14.

Mohan Dey, M., Rab, M. A., Paraguas, F. J., Bhatta, R., Ferdous Alam, M., Koeshendrajana, S., & Ahmed, M. (2005). Status and economics of freshwater aquaculture in selected countries of Asia. *Aquaculture Economics & Management*, *9*(1–2), 11–37.

Naik, R. S. (2017). Best management practices in aquaculture in India. *Best Management Practices in Aquaculture*, *21*.

Ngasotter, S., Panda, S. P., Mohanty, U., Akter, S., Mukherjee, S., Waikhom, D., & Devi, L. S. (2020). Current scenario of fisheries and aquaculture in India with special reference to Odisha: A review on its status, issues and prospects for sustainable development. *International Journal of Bio-resource and Stress Management*, *11*(Aug, 4), 370–380.

Suzuki, A. (2021). Rising importance of aquaculture in Asia: Current status, issues, and recommendations. *Asian Development Outlook Update Background Paper.* Asian Development Bank, Manila.

Tacon, A. G., & Metian, M. (2013). Fish matters: Importance of aquatic foods in human nutrition and global food supply. *Reviews in Fisheries Science*, *21*(1), 22–38.

Wu, C., Xiong, X., Hamidian, A. H., Zhang, Y., & Xu, X. (2022). A review on source, occurrence, and impacts of microplastics in freshwater aquaculture systems in China. *Water Biology and Security*, *1*(3), 100040.

Yadav, N. K., Patel, A. B., Singh, S. K., Mehta, N. K., Anand, V., Lal, J., Dekari, D. & Devi, N. C. (2024). Climate change effects on aquaculture production and its sustainable management through climate-resilient adaptation strategies: A review. *Environmental Science and Pollution Research*, *31*(22), 31731–31751.

CHAPTER 2

Minimizing Environmental Footprint: Toward Zero-Waste Aquaculture

SUPARNA DEB*, SHUBHAM KASHYAP, PRADYUT BISWAS, JHAMLAL, ANAMIKA DEBNATH, PAYEL DEBBARMA, and BHAVESH CHOUDHARY

College of Fisheries, CAU (I), Lembucherra, West Tripura, Tripura, India

*Corresponding author

ABSTRACT

Aquaculture, or fish farming, has the ability to provide a sustainable source of protein to a growing population. Traditional practices, however, can produce waste and have an influence on the environment. The notion of zero-waste aquaculture addresses this issue by eliminating waste products and establishing a closed-loop system. We can reduce our environmental footprint and achieve true zero-waste aquaculture by minimizing inputs such as fish feed and medications, using efficient feeding practices and promoting healthy fish through good water quality management, recycling water within the system through filtration and treatment, converting uneaten fish feed into compost for use in land-based agriculture systems, and implementing innovative techniques such as aquaponics, biofloc, and integrated multitrophic aquaculture. This zero-waste aquaculture offers several advantages, including reducing water pollution, conserving resources, and promoting a circular economy.

2.1 INTRODUCTION

Global growth of various food systems has sustained human well-being while also having a significant impact on environmental change (Willett et al., 2019). Aquaculture, like other farming activities, interacts

Advances in Freshwater Aquaculture: Technology for Sustainable Development. Tarang Kumar Shah, Madhu Sharma, Mohd. Ashraf Rather, & Avdhesh Kumar (Eds.)

DOI: 10.1201/9781779640857-2

with ecosystems involved in input production and usage (resource ecosystems) as well as those impacted by output release (receiving ecosystems). These interactions operate cumulatively and synergistically across different spatial scales (i.e., farm, land, marine, and globe) and can have both detrimental and positive consequences on ecosystems (Soto et al., 2007) (Figure 2.1). The variety of production technologies and stressors make it difficult to determine the aquaculture sector's overall environmental pressure. For starters, the vast majority of finfish and crustaceans are now being produced with high-quality compound feeds (FAO, 2018). Even herbivorous and omnivorous species such as carps, catfish, and tilapia, which might graze on the natural food web or agricultural leftovers in semi-intensive farming settings, are increasingly provided higher quality compound feeds (Edwards, 2015). This increased reliance on compound feed can increase local stressors on aquatic ecosystems (e.g., organic and inorganic nutrient emissions in the water) as well as distant stressors on land and aquatic ecosystems involved in feedstuff production (e.g., land conversion, pesticide and fertilizer use on land, and depletion of wild stocks) (Chary et al., 2023). Sustainable aquaculture production has significance for global food security due to human issues such as climate change, population expansion, and environmental degradation (Bank et al., 2020; Bush et al., 2013). The "safeguard" principle emphasizes the importance of protecting and restoring the health of agroecosystems (and thereby aquatic ecosystems). To protect ecosystem health, aquaculture must be kept within the ecosystem's carrying capacity. This necessitates regenerative systems and practices that do not impair ecosystem functioning or structure to irreversible or undesirable levels, or, better yet, can improve ecosystem service provision (Chary et al., 2023).

Environmental footprint, also known as ecological footprint, is a metric that assesses the impact of human activity on the biosphere. It represents the amount of biologically productive land and water area required to produce the resources we use and absorb the waste we make. There are numerous methods for calculating an environmental footprint, but all consider aspects such as land use, water usage, energy consumption, and climate change. Our environmental impact is always increasing, putting a pressure on the planet's resources. There are several things we can do to lessen our environmental impact such as like resource conservation and trash reduction.

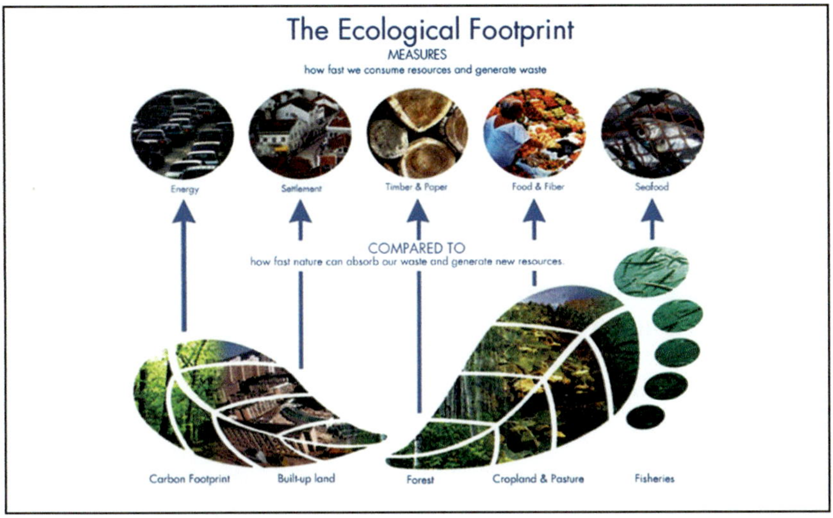

FIGURE 2.1 Ecological footprint.

Source: Global Footprint Network. https://www.footprintnetwork.org/content/uploads/2016/10/Footprint-highres.png

Extensive aquaculture can help with circularity by recycling nutrients; however the worldwide trend is toward more intensive production methods, and the enormous areas required for extensive production can quickly lead to land-use change. The second trend is the continued transition of protein supply from aquatic to crop-based elements for carnivorous finfish diets. This shift suggests that aquaculture's environmental impacts are increasingly affecting terrestrial ecosystems (e.g., soil degradation, deforestation) and freshwater resources (e.g., eutrophication, water depletion) (Cottrell et al., 2020; Pahlow et al., 2015). A better balance with aquaculture systems that use less water (e.g., recirculating aquaculture systems (RAS), or closed biofloc systems) and/or rely almost entirely on marine waters (e.g., seaweed, mussels) could reduce aquaculture's overall pressure on freshwater ecosystems (Chary et al., 2023).

Aquaculture is viewed as a double-edged sword since it provides enormous benefits to humans while also having significant environmental consequences (Smith et al., 2010). Aquatic goods have a high nutritional quality and are a significant source of macro- and micronutrients for people in underdeveloped countries (Roos et al., 2007). Additionally, aquaculture and fisheries are recognized as sources of employment (Smith et al., 2010).

2.2 DIFFERENT IMPACT OF AQUACULTURE ON ENVIRONMENTAL FOOTPRINT

Aquaculture has been accused of causing numerous environmental, social, and economic issues, as well as esthetic concerns. Ecosystems are not necessarily as fragile as one might think; instead, they have an incredible ability for resilience, and as long as essential processes are not irreversibly disrupted, ecosystems will continue to recycle and distribute energy (Frankic and Hershner, 2003). However, irrevocable damage has already occurred as a result of insufficient activity management (Figure 2.2).

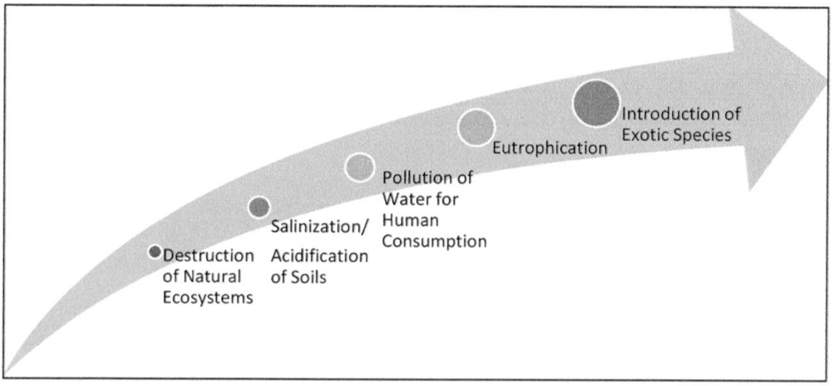

FIGURE 2.2 Impact of aquaculture on environmental footprint.

2.2.1 DESTRUCTION OF NATURAL ECOSYSTEMS TO CONSTRUCT AQUACULTURE FARMS

Tidwell and Allan (2001) cite mangrove forests as the primary source of organic matter in the coastal zone. They also serve as nursery habitats for many aquatic species that are ecologically and/or economically significant, as well as refuge or nesting areas for birds, reptiles, crabs, and other taxonomic groups. Mangroves also serve as accumulation places for sediments, pollutants, nitrogen, and carbon, as well as providing protection against coastal erosion (Alongi, 2002). The global cover of mangrove forest has dropped from 19.8 million hectares in 1980 to less than 15 million in 2000. The yearly deforestation rate was 1.7% from 1980 to 1990 and 1.0% from 1990 to 2000 (FAO, 2007), and the problem persists today. Some scholars have documented that aquaculture has caused the devastation of millions of hectares of mangrove forest in Thailand, Indonesia, Ecuador, Madagascar, and other countries (Naylor et al., 2000; Harper et al., 2007). According to Páez-Osuna (2001), the Philippines has converted 205,523 ha of mangrove and wetlands into aquaculture farms, whereas Indonesia has 211,000 ha, Vietnam 102,000 ha, Bangladesh 65,000 ha, and Ecuador 21,600 ha.

2.2.2 SALINIZATION/ACIDIFICATION

Aquaculture farms can be abandoned due to several reasons, including operational, economic, and hygienic issues, resulting in hypersaline, acidic, and degraded soil (Rodríguez-Valencia et al., 2010). As a result, those soils are unsuitable for agriculture and will remain unusable for an extended duration. Using lime and other aquaculture chemicals to remediate soil can alter its physicochemical properties, perhaps exacerbating the issue (Martínez-Córdova et al., 2009).

2.2.3 POLLUTION OF WATER BODY FOR HUMAN CONSUMPTION

According to Páez-Osuna (2001), inland aquaculture has negatively impacted human-consumed water sources. However, just a few research works have been undertaken on this topic.

2.2.4 EUTROPHICATION AND NUTRIFICATION OF EFFLUENT RECEIVING ECOSYSTEMS

The organic enrichment known as eutrophication is mainly produced by nonconsumed feed (especially due to overfeeding), lixiviation of aquaculture feedstuffs (Focardi et al., 2005), decomposition of died organisms, and overfertilization (Tacon and Forster, 2003; Feng et al., 2004). Current fish farming practices lead to significant pollution as a large portion of the nutrients added to feed the fish end up as waste. Studies show that only 20–50% of the added nitrogen is actually used by the fish themselves (Jackson et al., 2003; Schneider et al., 2005). The rest accumulates in the water or settles at the bottom (sediment). This excess nutrient pollution harms nearby ecosystems in various ways. It can trigger algal blooms, including harmful red tides (Alonso-Rodrıguez and Páez-Osuna, 2003). It can also smother and kill bottom-dwelling organisms (burying) and cause unpleasant odors and introduce pathogens into the surrounding environment (Córdova and Ocaña, 2007). The buildup of nutrients, like nitrogen (N), phosphorus (P), and carbon (C), in the water of aquaculture ponds is a major concern. This nutrification can happen in several ways, including adding fertilizer, organic matter decomposing, bottom sediments getting disturbed, and waste products from the fish themselves. The biggest worry is the rise in nitrogen-based waste, especially ammonia. In its uncharged form (NH_3), ammonia is highly toxic to many aquatic organisms, as research by Casillas-Hernández et al. (2007) has shown.

2.2.5 ECOLOGICAL IMPACTS CAUSED BY INADEQUATE MEDICATION PRACTICES

While fish farmers commonly use medications such as antibiotics and hormones to prevent disease and boost growth, these practices have come under scrutiny. Studies have found residues of various pharmaceuticals, including hormones, in surrounding water sources (Boxall, 2004). This raises concerns about potential environmental disruption, as these chemicals can cause imbalances in nearby ecosystems. The impact of hormone use in aquaculture, specifically, deserves further investigation.

2.2.6 OTHER ACCUSATIONS

- The agricultural and aquacultural activities have contributed to the degradation of ecosystems including important modification on landscape (DeWalt et al., 2000; Naylor et al., 2000).
- Some other accusations for aquaculture include the production of fish and shellfish with high concentrations of toxins and/or heavy metals; genetic pollution and infestation of nondesirable phytoplankton and/or zooplankton species (Johnson et al., 1996; Kruzynski, 2004).

To address the environmental issues of traditional aquaculture, zero-waste farming systems offer a compelling solution. These systems, like integrated multitrophic aquaculture (IMTA), integrated aquaponics, and biofloc technology, promote a circular economy by minimizing inputs, reusing resources, and responsibly recycling nutrient-rich water through plants.

This approach, as highlighted by de Korte et al. (2024), fosters water and nutrient conservation, making zero-waste aquaculture a sustainable path forward for fish farming (Figure 2.3).

2.3 SOME KEY PRINCIPLES OF ZERO-WASTE AQUACULTURE

2.3.1 REDUCE

Minimizing inputs like fish feed and medications by using efficient feeding practices and promoting healthy fish through good water quality management.

2.3.2 REUSE

Recycling water within the system through filtration and treatment. Uneaten fish feed or organic waste products from the fish can be converted into compost for use in land-based agriculture.

2.3.3 RECYCLE

Using innovative techniques like aquaponics, which combines aquaculture with hydroponic plant cultivation. The nutrient-rich water from the fish tanks fertilizes the plants, while the plants help filter the water for the fish.

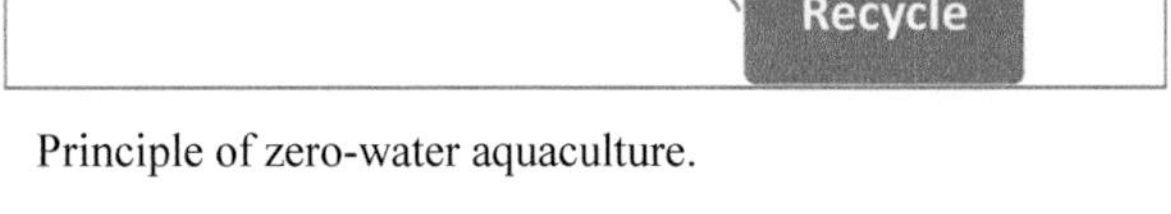

FIGURE 2.3 Principle of zero-water aquaculture.

Studies highlight that waste from fish and shrimp farms, if not managed effectively, pollutes aquatic environments. This waste includes uneaten food and fish/shrimp waste that sink and accumulate on the seabed. As research by Andhikawati and Permana (2023) suggests, this can lead to significant water pollution due to high levels of nitrogen (from feed) and phosphorus accumulating in the water.

Driven by factors such as resource scarcity, rising energy costs, and stricter environmental regulations, the aquaculture industry is looking for ways to improve production practices. One promising approach is the ecoindustrial park concept, which promotes a closed-loop system where different industries work together. In this model, each "node" focuses on food production or waste conversion, creating a highly self-sufficient system that addresses current economic and environmental challenges (Figure 2.5).

In Figure 2.4, green dashed box represents the RAS, the red dashed box the hydroponic production system (HPS), and the orange dashed box the anaerobic digestion system.

2.4 UTILIZATION OF AQUACULTURE WASTE

Some of the examples of how to utilize aquaculture waste are given below.

- Wastewater from catfish farms can be utilized as liquid organic fertilizer. This is because waste still includes nutrients such as nitrogen (N) and phosphorus (F), which can be used as liquid organic fertilizer (Hope, 2019).

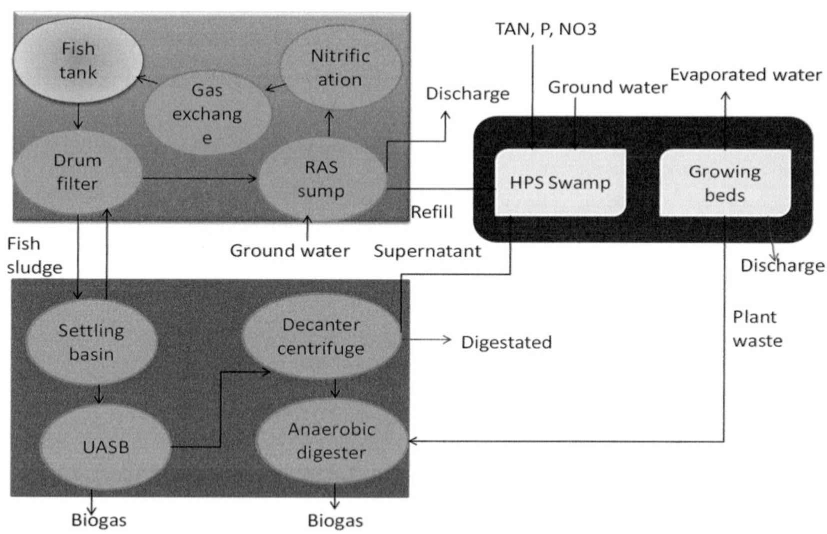

FIGURE 2.4 Ecoindustrial food park.

- Waste from fish farms can be used to raise silkworms. This is because prawn waste in the form of feces still contains nutritional value in the form of protein and has small particle sizes of fish feces that are appropriate for silkworm rearing media. As a result, the use of fish farming waste, specifically fish feces, will have an impact on the growth of these silkworms (Ardana et al., 2018).

- Organic waste from catfish farming operations can be used as an aquaponic medium. This is because fish farming waste still includes nutrients that can be used to cultivate plants and vegetables (Andhikawati and Permana, 2023).
- Liquid waste from prawn aquaculture can be used as an alternative medium for microalgae development. This is because shrimp wastewater still contains approximately 77% nitrogen and 85% phosphorus, making it suitable as a growth medium for *Chaetoceros amami* microalgae (Andhikawati and Permana, 2023).
- Waste from goldfish aquaculture can be utilized as fish food. This is because fish farming waste consists of organic waste in the form of residual feed and fish feces, which can be utilized as a medium for producing plankton microorganisms that can be used as a fish food source (Makhfirah et al., 2018).
- *Spirulina platensis* can be maintained using liquid waste in the form of water from catfish farms. This is because wastewater from fish aquaculture still includes nitrogen, ammonia, and carbon dioxide, which can promote the growth of *S. platensis* (Widyantoro et al., 2018).
- Water waste from catfish farming activities is employed as a medium for sustaining *Daphnia* sp. This is because the wastewater still contains organic substance in the form of N and NH_3 from leftover feed and catfish metabolic waste, which can be used as a protein source for *Daphnia* sp. growth (Akmal et al., 2019).

2.5 PRINCIPLE OF AQUAPONICS

Aquaponics combines fish farming (RAS) with plant cultivation (HPS) to create a closed-loop system. Fish waste provides nutrients for the plants, reducing reliance on external resources and minimizing waste disposal (Goddek et al., 2018; Junge et al., 2017; vanWoensel, 2015). Traditionally, all components (fish, plants, bacteria) share the same water that is constantly circulated (Naegel, 1977; Rakocy, 2012). However, recent research explores separating the systems to better meet the specific needs of each species (Baganz et al., 2022; Goddek et al., 2016; Kloas et al., 2015; Yep and Zhang, 2019). Aquaponics requires a deep understanding of how water, nutrients, and energy flow between the different elements (Greenfeld et al., 2020). This resembles a miniature ecoindustrial park (Côté and Cohen-Rosenthal, 1998; Lowe et al., 1997) where resources are efficiently used, making it a great fit for the circular economy with its economic, environmental, and social benefits (Hein et al., 2015).

2.6 PRINCIPLE OF IMTA

With the expansion of the fisheries industry, new cultural practices and improvements to culture systems for the blue revolution have been implemented. One such system is IMTA. IMTA is an intense and synergistic cultivation method that employs water-borne nutrients and energy transfer between multiple species at different trophic levels. It uses the outputs, including waste, of one aquatic species as input (fertilizer, food) for another.

IMTA supports economic and environmental sustainability by transforming wastes and uneaten feed from fed organisms into harvestable crops, reducing waste and enhancing economic diversity. Many projects on IMTA have been completed in various countries. Aquaculture allows for the employment of many species from various trophic levels, which is incredibly productive. IMTA has the potential to significantly accelerate the blue revolution in the near future (Lal et al., 2023).

2.7 PRINCIPLE OF BIOFLOC

Traditional aquaculture operations are being innovated using a variety of modern technology like biofloc. Biofloc increases bacteria's heterotrophic capacity, reduces water use, and boosts nutrient recycling. Biofloc technology has the ability to boost the growth and health of fish species by increasing water quality. Furthermore, it emphasizes biofloc's economic and ecological benefits, as well as its capacity to address the issues of aquaculture techniques. Biofloc subsidizes the feed conversion ratio, lowering the cost. These activities supply minerals, nutrients, and fatty acids to the rearing species all day. Furthermore, biofloc governs sustainable methods that ensure great profitability when compared with traditional aquaculture procedures. Finally, biofloc technology demonstrates excellent outcomes through minimum water exchange, improved water quality, and lower production costs, resulting in a high profitability ratio when compared with local fishery techniques (Kobir et al., 2024).

2.8 ENVIRONMENTAL BENEFITS OF IMPLEMENTING STRATEGIES IN ZERO-WASTE AQUACULTURE

- **Reduced water pollution:** Decreases nutrient runoff from fish waste into surrounding water bodies.
- **Conservation of resources:** Minimizes the need for fresh water and fish feed.
- **Promoting circular economy:** Creates a closed-loop system where waste from one process becomes a resource for another (Figure 2.5).

While achieving true zero-waste is challenging, focusing on these principles can significantly reduce the environmental impact of aquaculture. Here are some areas for further exploration:

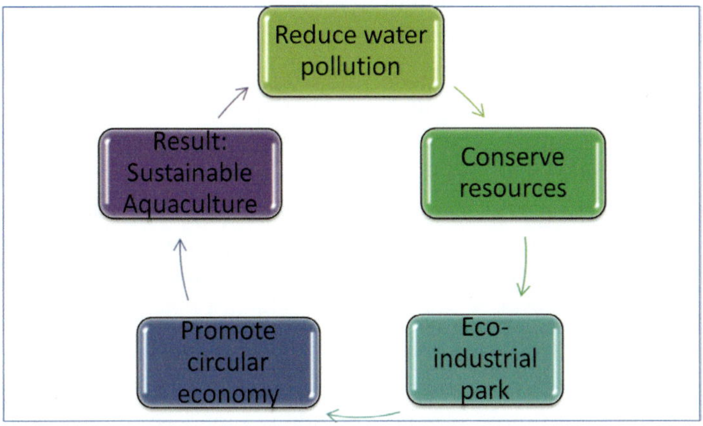

FIGURE 2.5 Environmental and other benefits of zero-water aquaculture.

- **Development of new feed sources:** Research on alternative, sustainable feed ingredients like insect meal or algae can lessen the reliance on fishmeal, which can contribute to overfishing.
- **Improved waste treatment technologies:** Advancements in waste conversion can turn byproducts from fish farming into valuable products like biofuels or fertilizers.

By embracing innovation and adopting a circular economy approach, zero-waste aquaculture can become a reality, ensuring a sustainable and responsible future for fish farming.

2.9 CIRCULAR ECONOMY

A circular economy is thought to be one strategy to lessen the environmental impacts of human activities by using resources more efficiently and recovering them, resulting in less waste and emissions than linear take-make-dispose systems. The circular economy is the use of solid waste from fishing and the collecting of these species in extractive zones, in accordance with the United Nations Sustainable Development Goals. Animal feed, organic fertilizer, biofuels, crab shell chitin biocomposite derivatives, and chitosan-based composites for food packaging applications are the most frequently mentioned byproducts due to their nontoxicity, antibacterial, and antifungal qualities (Filho et al., 2024). Muscat et al. (2021) proposed five ecological principles to guide biomass use toward a circular economy.

On these principles, the author gives a narrative overview to (1) apply them to aquaculture while outlining consequences for the main species and production systems, and (2) identify the primary approaches to making aquaculture more circular. We demonstrate that the basic concepts of "safeguard," "entropy," and "recycle" principles have been thoroughly explored and, in certain cases, effectively implemented. In contrast, the "avoid" and "prioritize" concepts have received much less attention; doing so would provide a chance to reduce the environmental implications of aquaculture at the food-system level. One example is prioritizing the production of species that contribute to food and nutrition security, have low environmental consequences, and consider the larger food system scale to avoid feed-food competition in aquaculture. Six priorities for making aquaculture more circular include (1) increasing production and demand for the most important species, (2) reducing food loss and waste at farm and post-harvest stages, (3) supporting nutrient recycling practices at multiple scales, (4) adapting aquafeed formulations, (5) informing consumers about the benefits of low-trophic-level species and other environmentally friendly aquatic foods, and (6) addressing urgent research gaps (Chary et al., 2024).

Climate change, droughts, increased demand, population expansion, and poor water management practices are all contributing to the growing scarcity of freshwater. Salination of aquaculture is a key issue, particularly in coastal or backwater aquaculture. Biodesalination with microalgae is a viable long-term answer to this dilemma. Mirzaei et al. (2024) show that

Arthrospira platensis and *Dunaliella salina* may thrive in saline aquaculture wastewater, lowering salinity and nutrient concentrations. Field emission scanning electron microscopy revealed the creation of a salt film on the microalgae's surface as a result of salt biosorption and accumulation. Fourier transform infrared spectroscopy investigation revealed the binding of functional groups inside the cell wall of *A. platensis* and the cell membrane of *D. salina* to the ions in the medium. Scaling up the development of *A. platensis* in a photobioreactor under nonsterile circumstances demonstrated the potential of the processes for industrial-scale biodesalination. Although the treated water did not meet irrigation or drinking water standards, this strategy provides a preliminary desalination stage, lowering the strain on following treatments while also removing nutrients and producing biomass (Mirzaei et al., 2024).

The livestock sector is an important part of agriculture and contributes significantly to global food security. Animal production is critical to meeting future demand for animal products as the world's population grows and improving people's socioeconomic position, particularly in developing countries. For this, it is critical to identify and use novel feed resources that are less competitive with human foods, contribute to reducing the environmental footprint of the livestock sector, and can help cope with the negative consequences of climate change in future livestock production. The proper use of alternative and environmentally friendly land-based and aquatic resources can play an important role in building sustainable and efficient livestock production in the future (Khanal, 2024). By utilizing a variety of alternative resources, with a focus on insects, marine macroalgae, and freshwater aquatic plants and algae, we can increase the capacity of alternative feed resources, which could play an important role in achieving a circular economy in livestock production through education and research activities.

Grape byproducts have already been employed in cosmetics and food industries, as well as animal feed, particularly monogastrics and aquaculture, due to their antibacterial and antioxidant properties. In aquaculture, the high demand and need to substitute animal feeds with vegetable sources has positioned grape byproducts as potential new phytonutrients with therapeutic qualities. Although other studies have found that high supplementation of grape byproducts can have a negative impact on fish health and growth due to antinutritional factors (tannins), grape byproducts have been shown to be valuable phytonutrients that can be incorporated into fish feed to improve growth and health during rearing conditions and boost the sustainable circular economy by increasing the use of these products (Quagliardi et al., 2024).

Marine plastic pollution has emerged as a major serious concern in the 21st century, threatening planetary boundaries, biological, and anthropogenic systems. Such complex and multiscale concerns have a key role in the degradation of our land and ocean ecosystems, for example, loss of aquatic life as a result of ghost fishing or microplastic pollution from broken-down plastics in oceans. These "wicked societal concerns" can no longer be addressed using our current society institutions and procedures. The aquaculture and fishing industries are major producers of plastic marine litter worldwide. To achieve more sustainable circular practices in the use and management of plastic in the fisheries and aquaculture sectors, interdisciplinary cross-sector collaboration is required, which addresses such global concerns at the local level. The systems design approach and framework have created collaborative spaces and interactions that can generate knowledge, insights, tool creation, and creativity in the quest of innovation and circular transitions for these wicked societal issues (Palmer-Abbs et al., 2024).

2.10 TOWARD BLUE ECONOMY

Fishery, as an economic activity, plays an essential part in the country's progress and social welfare. Fisheries, as an economic activity, should be promoted, developed, and prioritized to contribute to good economic growth. This necessitates innovation and ingenuity, as well as optimizing the role of stakeholders, in this case the government, to ensure that fisheries continue to exist and contribute to the country's welfare. Blue economy is a concept that focuses on optimizing aquatic resources to boost economic growth through a variety of innovative and creative activities while assuring company continuity and environmental sustainability. Blue economy prioritizes and emphasizes efficiency. Efficiency promotes investment and fishing business development while also ensuring a sustainable environment. Blue economy is built on the comprehensive growth of the people's economy, with the goal of dramatically increasing national development. Blue economy's major focus is on operations that benefit the ecosystem. Waste, both chemical and organic waste, will have a direct or indirect impact on habitat and ecosystem life; thus, science and technology must be used to treat waste output.

Gunter Pauli was the first to use the term blue economy in his 2010 book *The Blue Economy: 10 Years, 100 Innovations, 100 Million Jobs*. This blue economy concept is based on ecosystem logic, which states that ecosystems strive for greater efficiency in the flow of nutrients and energy to meet the basic needs of all contributors in a system (Chandra et al., 2021). Blue economy also emphasizes innovation and creativity, which include product diversity, production system efficiency, and resource management system structure (Andhikawati and Permana, 2023).

Blue economy is viewed as an alternative possibility to address environmental issues and poverty that are prevalent in coastal towns. The concept of blue economy is projected to boost the marine and fisheries economic sectors. Furthermore, blue economy emphasizes the notion of zero waste, which ensures that every natural resource is used optimally by humans (Nurfauzi, 2022). Organic matter left over from fish feed, the use of drugs (chemotherapeutic agents) to treat fish diseases, and the use of other chemicals, such as disinfectants to treat fish culture media before use, all contribute to water pollution in the small-, medium-, and large-scale aquaculture industries. To address this, academics and researchers have developed a number of technologies, including the use of natural or probiotic bacteria to treat fish diseases and cultivation media, as well as the use of IMTA technology, which eliminates waste from the remaining fish feed. All waste generated in the aquaculture, catch fisheries, and processing industries has received little attention from commercial actors. The emergence of blue economy, launched by the Ministry of Maritime Affairs and Fisheries, is one of the concepts used to emphasize and remind the importance of managing waste products from fishing activities so that they do not pollute the environment and thus the environmental ecosystem is preserved (Figure 2.6) (Andhikawati and Permana, 2023).

2.11 WASTE MANAGEMENT STRATEGIES

2.11.1 CARRY OUT PERIODIC MONITORING OF INTERNAL AND EXTERNAL EVALUATIONS OF WASTE MANAGEMENT

Monitoring and evaluating waste management is critical for determining its performance and the environmental conditions of the surrounding waters.

FIGURE 2.6 Blue economy.

Internal monitoring is carried out by cultivation owners or enterprises using environmental management records in line with relevant rules and regulations (Jolly et al., 2023). Basically, waste monitoring must be implemented whenever garbage is dumped into the seas. So, for aquaculture activities that operate on a daily basis and generate waste, daily waste quality monitoring is essential to detect oscillations in effluent quality and guarantee that the quality satisfies the stated standards. Daily effluent quality monitoring can be accomplished through real-time monitoring of some parameters *in situ* such as temperature, pH, brightness, turbidity, dissolved oxygen concentration, and effluent discharge. External monitoring is carried out by authorized agencies involved in cultivation. The significance of internal and external monitoring can serve as control papers in attempts to maintain a sustainable environment.

2.11.2 FORMULATE STRATEGIC INTERSECTORAL POLICIES FOR THE CENTRAL/REGIONAL GOVERNMENT ON STANDARD MANAGEMENT SYSTEMS AND QUALITY OF AQUACULTURE WASTE

Strategic intersectoral policies are required to manage aquaculture waste. Currently, there is no clear regulatory reference for binding or required aquaculture waste effluent quality requirements established by the government.

Environmental issues necessitate cross-sectoral policies or the implementation of a collaborative cooperation framework between government agencies responsible for the environment and those in charge of the fishing industry. The development of this strategy is critical in raising aquaculture entrepreneurs' knowledge of environmental issues. It is hoped that this plan will influence the regularity with which all parties involved support and ensure the quality of trash generated by intense aquaculture activities. It is crucial to note that when the government or an organization establishes a permit policy and waste criteria, it rarely includes a request for procedures to be followed in order to comply with the policy. If the waste does not meet permits or regulations, methods to increase waste quality must be included, as well as management adjustments to produce better quality trash (Boyd, 2003).

2.11.3 INCREASE CROSS-SECTOR COLLABORATION BETWEEN THE CENTRAL/LOCAL GOVERNMENT ON SUSTAINABLE CULTIVATION SYSTEM POLICIES

Sustainable farming is based on economic, ecological, and social balance. This can be achieved through cross-sector coordination between central and state governments, which have control over all aspects of sustainability in policy development. The primary goals of designing a sustainable agriculture system are to maximize economic potential, create a sustainable natural environment, and improve social welfare. Apart from that, aquaculture entrepreneurs must support and adhere to government policies that promote sustainable production beyond economic rewards. One alternate approach to achieving sustainable aquaculture activities is to implement or adopt the best management practices (BMPs) system. According to Boyd (2003), BMPs are international terms for environmental management actions carried out to reduce water pollution and other negative environmental impacts.

2.12 STRATEGY FOR UTILIZING AQUACULTURE WASTE

The technical steps of strategy formulation (strengths, weaknesses, opportunities, and threats (SWOT) analysis) refer to the process for utilizing aquaculture waste as a renewable resource with the zero-waste philosophy. A strategic analysis must consider a variety of elements before selecting a specific alternative that is seen to be the best decision. The SWOT analysis is a tool for evaluating many elements that are important to consider. SWOT analysis is the systematic identification of various aspects to develop a strategy. This analysis is based on logic that maximizes strengths and opportunities while minimizing weaknesses and risks. SWOT analysis is used to analyze strategies for utilizing aquaculture waste (Figure 2.7).

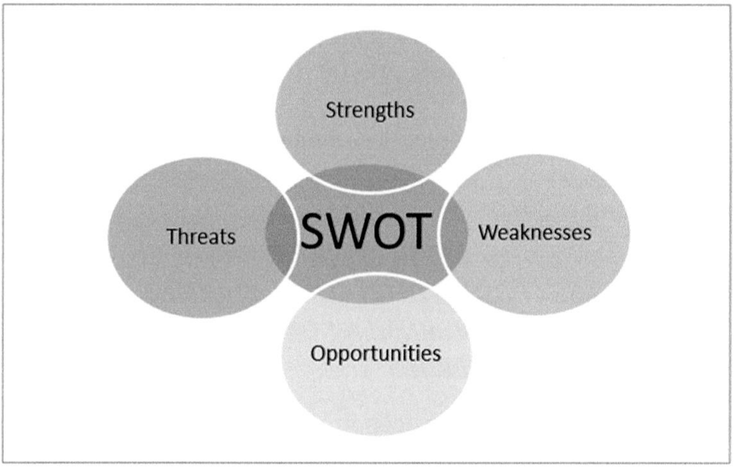

FIGURE 2.7 SWOT analysis.

2.12.1 STRENGTHS

Aquaculture waste is heavy in nutrients, and the residual feed produces sediment waste, which is made up of organic and inorganic components. The precipitated sludge particles can be used as fertilizer because they contain high levels of N, P, and K, allowing aquaculture waste to be reused and becoming a distinct advantage over aquaculture systems that previously produced waste has now become zero waste, making this strategy suitable for utilizing aquaculture waste.

2.12.2 WEAKNESSES

The problem in the utilization of aquaculture waste is a lack of information among fish farming industry players, which results in aquaculture waste being thrown away, left alone, or channeled straight into the environment of the waste receiving agency.

2.12.3 OPPORTUNITIES

Despite the shortcomings stated above, aquaculture waste has a tremendous potential when used. When this trash is repurposed, it improves soil fertility. There are numerous sorts of natural food in the pond that are essential for the growth of shellfish and finfish.

Klekap, for example, is a collection of microorganisms that live on the pond bottom's surface, including blue algae, benthos, diatoms, bacteria, and animal microorganisms. Klekap creates a sheet of brown, greenish brown, yellowish green, or blue green, depending on the type and proportion of algae present (Andhikawati and Permana, 2023).

2.12.4 THREATS

The issue that growers are most concerned about when using the high nutrients obtained from aquaculture waste is the emergence of excessive fertility (eutrophication), which would result in an explosion of phytoplankton in the waterways, ultimately disrupting the existing ecosystem.

2.13 CONCLUSION

To summarize, zero-waste aquaculture is a viable approach for sustainable fish production. This strategy decreases environmental effect while providing a consistent source of protein by minimizing waste and implementing closed-loop technologies. Zero-waste aquaculture emphasizes water conservation, resource efficiency, and a circular economy through efficient practices, water treatment, and novel technology such as aquaponics and biofloc, positioning it as a responsible and sustainable path for fish farming in the future.

KEYWORDS

- **zero-waste aquaculture**
- **environmental footprint**
- **circular economy**
- **integrated multitrophic aquaculture**

REFERENCES

Agriculture Organization of the United Nations. (2018). Fisheries department. In: *The State of World Fisheries and Aquaculture*. Food and Agriculture Organization of the United Nations.

Akmal, Y., Humairani, R., & Zulfahmi, I. (2019). Utilization of Catfish (*Clarias* sp.) Wastewater as a Cultivation Media for Daphnia sp. *Journal of Biosciences and Education*, *1*, 22–27.

Alongi, D. M. (2002). Present state and future of the world's mangrove forests. *Environmental Conservation*, *29*(3), 331–349.

Alonso-Rodrıguez, R., & Páez-Osuna, F. (2003). Nutrients, phytoplankton and harmful algal blooms in shrimp ponds: A review with special reference to the situation in the Gulf of California. *Aquaculture*, *219*(1–4), 317–336.

Andhikawati, A., & Permana, R. (2023). Management of liquid and solid waste from aquaculture in Indonesia. *International Journal of All Research Writings*, *4*(8), 1–8.

Ardana, P., Abidin, Z., & Diniarti, N. (2018). Utilization of fish cultivation to increase the growth of silk worm biomass (*Tubifex* sp.). *Journal of Fisheries*, *8*(1), 55–64.

Baganz, G. F., Junge, R., Portella, M. C., Goddek, S., Keesman, K. J., Baganz, D., Staaks, G., Shaw, C., Lohrberg, F., & Kloas, W. (2022). The aquaponic principle—It is all about coupling. *Reviews in Aquaculture*, *14*(1), 252–264.

Bank, M. S., Metian, M., & Swarzenski, P. W. (2020). Defining seafood safety in the Anthropocene. *Environmental Science & Technology*, *54*(14), 8506–8508.

Boxall, A. B. (2004). The environmental side effects of medication: How are human and veterinary medicines in soils and water bodies affecting human and environmental health. *EMBO Reports*, *5*(12), 1110–1116.

Boyd, C. E. (2003). Guidelines for aquaculture effluent management at the farm-level. *Aquaculture*, *226*(1–4), 101–112.

Bush, S. R., Belton, B., Hall, D., Vandergeest, P., Murray, F. J., Ponte, S., Oosterveer, P., Islam, M. S., Mol, A. P. J., Hatanaka, M., Kruijssen, F., Ha, T. T. T., Little, D. C., & Kusumawati, R. (2013). Certify sustainable aquaculture. *Science*, *341*(6150), 1067–1068.

Casillas-Hernández, R., Nolasco-Soria, H., García-Galano, T., Carrillo-Farnes, O., & Páez-Osuna, F. (2007). Water quality, chemical fluxes and production in semi-intensive Pacific white shrimp (*Litopenaeus vannamei*) culture ponds utilizing two different feeding strategies. *Aquacultural Engineering*, *36*(2), 105–114.

Chandra, Yes, Rustam, I., & Safitri, P. (2021). Implementation of blue economy-based policies within the framework of Indonesian government cooperation with FAO: A study of fish processing units in north Lombok regency. *Indonesian Journal of Global Discourse*, *3*(1), 1–19.

Chary, K., van Riel, A. J., Muscat, A., Wilfart, A., Harchaoui, S., Verdegem, M., Filgueira, R., Troell, M., Henriksson, P. J. G., de Boer, I. J. M., & Wiegertjes, G. F. (2023). Transforming sustainable aquaculture by applying circularity principles. *Reviews in Aquaculture*, *16*(2), 656–673.

Chary, K., van Riel, A. J., Muscat, A., Wilfart, A., Harchaoui, S., Verdegem, M., Filgueira, R., Troell, M., Henriksson, P. J. G., de Boer, I. J. M., & Wiegertjes, G. F. (2024). Transforming sustainable aquaculture by applying circularity principles. *Reviews in Aquaculture*, *16*(2), 656–673.

Córdova, L. R. M., & Ocaña, L. F. E. (2007). Study of benthic fauna in the discharge lagoon of a shrimp farm with special emphasis on the polychaetes. *Online Journal of Biological Sciences*, *7*, 12–17.

Côté, R. P., & Cohen-Rosenthal, E. (1998). Designing eco-industrial parks: A synthesis of some experiences. *Journal of Cleaner Production*, *6*(3–4), 181–188.

Cottrell, R. S., Blanchard, J. L., Halpern, B. S., Metian, M., & Froehlich, H. E. (2020). Global adoption of novel aquaculture feeds could substantially reduce forage fish demand by 2030. *Nature Food*, *1*(5), 301–308.

de Korte, M., Bergman, J., van Willigenburg, G., & Keesman, K. J. (2024). Towards a zero-waste aquaponics-centered eco-industrial food park. *Journal of Cleaner Production*, *454*, 142109.

DeWalt, B. R., Zavala, J. R. R., Noriega, L., & González, R. E. (2000). *Shrimp Aquaculture, The People and the Environment in Coastal Mexico*. World Bank, NACA, WWF and FAO Consortium Program on Shrimp Farming and the Environment.

Edwards, P. (2015). Aquaculture environment interactions: Past, present and likely future trends. *Aquaculture*, *447*, 2–14.

FAO. (2007). *El Estado Mundial de la Pesca y Acuicultura 2006*. FAO: Rome, Italy.

Feng, Y. Y., Hou, L. C., Ping, N. X., Ling, T. D., & Kyo, C. I. (2004). Development of mariculture and its impacts in Chinese coastal waters. *Reviews in Fish Biology and Fisheries*, *14*, 1–10.

Filho, J. J. D. S., Gaspar, P. D., Souza, A. C. D., & Paço, A. D. (2024). Circular economy in guaiamum and Uçá crab waste in Brazil: Potential by-products—A systematic literature review. *Resources*, *13*(3), 46.

Focardi, S., Corsi, I., & Franchi, E. (2005). Safety issues and sustainable development of European aquaculture: New tools for environmentally sound aquaculture. *Aquaculture International*, *13*, 3–17.

Frankic, A., & Hershner, C. (2003). Sustainable aquaculture: Developing the promise of aquaculture. *Aquaculture International*, *11*, 517–530.

Goddek, S., Delaide, B. P., Joyce, A., Wuertz, S., Jijakli, M. H., Gross, A., Eding, E. H., Bläser, I., Reuter, M., Paul Keizer, L. C., Morgenstern, R., Körner, O., Verreth, J., & Keesman, K. J. (2018). Nutrient mineralization and organic matter reduction performance of RAS-based sludge in sequential UASB-EGSB reactors. *Aquacultural Engineering*, *83*, 10–19.

Goddek, S., Espinal, C. A., Delaide, B., Jijakli, M. H., Schmautz, Z., Wuertz, S., & Keesman, K. J. (2016). Navigating towards decoupled aquaponic systems: A system dynamics design approach. *Water*, *8*(7), 303.

Greenfeld, A., Becker, N., Bornman, J. F., & Angel, D. L. (2020). Identifying knowledge levels of aquaponics adopters. *Environmental Science and Pollution Research*, *27*, 4536–4540.

Harper, G. J., Steininger, M. K., Tucker, C. J., Juhn, D., & Hawkins, F. (2007). Fifty years of deforestation and forest fragmentation in Madagascar. *Environmental Conservation*, *34*(4), 325–333.

Hein, A. M., Jankovic, M., Farel, R., & Yannou, B. (2015, August). A conceptual framework for eco-industrial parks. In *International Design Engineering Technical Conferences and Computers and Information in Engineering Conference* (Vol. 57113, p. V004T05A024). American Society of Mechanical Engineers.

Hope, K. (2019). Utilization of nitrogen elements from catfish cultivation wastes as liquid fertilizer. *Journal of Education and Development*, *7*, 80–83

Jackson, C., Preston, N., Thompson, P. J., & Burford, M. (2003). Nitrogen budget and effluent nitrogen components at an intensive shrimp farm. *Aquaculture*, *218*(1–4), 397–411.

Johnson, S. C., Blaylock, R. B., Elphick, J., & Hyatt, K. D. (1996). Disease induced by the sea louse (*Lepeophteirus salmonis*) (Copepoda: Caligidae) in wild sockeye salmon (*Oncorhynchus nerka*) stocks of Alberni Inlet, British Columbia. *Canadian Journal of Fisheries and Aquatic Sciences*, *53*(12), 2888–2897.

Jolly, C. M., Nyandat, B., Yang, Z., Ridler, N., Matias, F., Zhang, Z., Murekezi, P., & Menezes, A. (2023). Dynamics of aquaculture governance. *Journal of the World Aquaculture Society*, *54*(2), 427–481.

Junge, R., König, B., Villarroel, M., Komives, T., & Jijakli, M. H. (2017). Strategic points in aquaponics. *Water*, *9*(3), 182.

Khanal, P. (2024). Use of land-based and aquatic alternative feed resources to establish a circular economy within livestock production. *Journal of Agriculture and Food Research*, *16*, 101087.

Kloas, W., Groß, R., Baganz, D., Graupner, J., Monsees, H., Schmidt, U., Staaks, Georg B. O., Suhl, J., Tschirner, M., Wittstock, B., Wuertz, S., Zikova, A., & Rennert, B. (2015). A new concept for aquaponic systems to improve sustainability, increase productivity, and reduce environmental impacts. *Aquaculture Environment Interactions*, *7*(2), 179–192.

Kobir, M. H., Hossain, M. A., Hossain, M. A., Dey, S. K., & Mithun, M. H. (2024). Biofloc based farming of Nile tilapia (*Oreochromis niloticus*) in tanks under different stocking densities. *Archives of Agriculture and Environmental Science*, *9*(1), 118–125.

Kruzynski, G. M. (2004). Cadmium in oysters and scallops: The BC experience. *Toxicology Letters*, *148*(3), 159–169.

Lal, J., Singh, S. K., Pawar, L., Biswas, P., Meitei, M. M., & Meena, D. K. (2023). Integrated multi-trophic aquaculture: A balanced ecosystem approach to blue revolution. In *Advances in Resting-state Functional MRI*. Woodhead Publishing, pp. 513–535.

Lowe, E. A., Warren, J. L., & Moran, S. R. (1997). *Discovering Industrial Ecology: An Executive Briefing and Sourcebook*. Battelle Press.

Makhfirah, H., Defira, Cn, & Hasri, I. (2018). Utilization of goldfish (*Cyprinus carpio*) cultivation waste on growth and survival of pressed fish (*Osteochilus kappeni*) with different stocking densities. *Unsiyah Maritime and Fisheries Student Scientific Journal*, *3*(2), 55–65.

Martínez-Córdova, L. R., Martínez-Porchas, M., & Pedrín-Avilés, S. (2009). Selección de sitios, construcción y preparación de estanques. *Camaronicultura Sustentable. chapter IDF, Mexico: Trillas*, p. 179.

Mirzaei, M., Jazini, M., Aminiershad, G., & Refardt, D. (2024). Biodesalination of saline aquaculture wastewater with simultaneous nutrient removal and biomass production using the microalgae Arthrospira and Dunaliella in a circular economy approach. *Desalination*, *581*, 117564.

Muscat, A., de Olde, E. M., Ripoll-Bosch, R., Van Zanten, H. H., Metze, T. A., Termeer, C. J., van Ittersum, M. K., & de Boer, I. J. (2021). Principles, drivers and opportunities of a circular bioeconomy. *Nature Food*, *2*(8), 561–566.

Naegel, L. C. (1977). Combined production of fish and plants in recirculating water. *Aquaculture*, *10*(1), 17–24.

Naylor, R. L., Goldburg, R. J., Primavera, J. H., Kautsky, N., Beveridge, M. C., Clay, J., Folke, C., Lubchenco, J., Mooney, H., & Troell, M. (2000). Effect of aquaculture on world fish supplies. *Nature*, *405*(6790), 1017–1024.

Nurfauzi, N. (2022). Implementation of the blue economy concept in coastal community development in the new normal period. *National Seminar on Economics and Accounting*, *1*(1), 75–82.

Páez-Osuna, F. (2001). The environmental impact of shrimp aquaculture: Causes, effects, and mitigating alternatives. *Environmental Management*, *28*, 131–140.

Pahlow, M., Van Oel, P. R., Mekonnen, M. M., & Hoekstra, A. Y. (2015). Increasing pressure on freshwater resources due to terrestrial feed ingredients for aquaculture production. *Science of the Total Environment*, *536*, 847–857.

Palmer-Abbs, M., Deshpande, P., & Karl, C. W. (2024). Enabling transition thinking on complex issues (wicked problems): A framework for future circular economic transitions of plastic management in the Norwegian fisheries and aquaculture sectors. *Journal of Cleaner Production*, *449*, 141420.

Quagliardi, M., Frapiccini, E., Marini, M., Panfili, M., Santanatoglia, A., Nguefang, M. L. K., Roncarati, A., Vittori, S., & Borsetta, G. (2024). Use of grape by-products in aquaculture: New frontiers for a circular economy application. *Heliyon*, *10*(5), e27443.

Rakocy, J. E. (2012). Aquaponics—integrating fish and plant culture. In: *Aquaculture Production Systems*, pp. 344–386.

Rodríguez-Valencia, J. A., Crespo, D., & López-Camacho, M. (2010). La camaronicultura y la sustentabilidad del Golfo de California. *WWF-México, ProgramaGolfo de California*.

Roos, N., Wahab, M. A., Chamnan, C., & Thilsted, S. H. (2007). The role of fish in food-based strategies to combat vitamin A and mineral deficiencies in developing countries. *The Journal of Nutrition*, *137*(4), 1106–1109.

Schneider, O., Sereti, V., Eding, E. H., & Verreth, J. A. J. (2005). Analysis of nutrient flows in integrated intensive aquaculture systems. *Aquacultural Engineering*, *32*(3–4), 379–401.

Smith, M. D., Roheim, C. A., Crowder, L. B., Halpern, B. S., Turnipseed, M., Anderson, J. L., Asche, F., Bourillón, L., Guttormsen, A. G., & Selkoe, K. A. (2010). Sustainability and global seafood. *Science*, *327*(5967), 784–786.

Soto, D., Aguilar-Manjarrez, J., Brugère, C., Angel, D., Bailey, C., Black, K., Edwards, P., Costa-Pierce, B. A., Chopin, T., Deudero, S., Freeman, S., Hambrey, J., Hishamunda, N., Knowler, D., Silvert, W., Marba, N., Mathé, S., Norambuena, R., Simard, F., Tett, P., Troell, M., & Wainberg, A. (2007, May). Applying an ecosystem-based approach to aquaculture: Principles, scales and some management measures. In *Building an Ecosystem Approach to Aquaculture. FAO/Universitat de les IllesBalears Expert Workshop* (Vol. 7, p. e11).

Tacon, A. G., & Forster, I. P. (2003). Aquafeeds and the environment: Policy implications. *Aquaculture*, *226*(1–4), 181–189.

Tidwell, J. H., & Allan, G. L. (2001). Fish as food: Aquaculture's contribution. *EMBO Reports*, *2*, 958–963.

vanWoensel, L. (2015). Ten technologies which could change our lives: Potential impacts and policy implications.

Widyantoro, H., Wijayanti, M., & Dwinanti, S. H. (2018). Modification of Spirulina platensis media as an effort to utilize catfish farming wastewater. *Jurnal Akuakultur Rawa Indonesia*, *6*(2), 153–164.

Willett, W., Rockström, J., Loken, B., Springmann, M., Lang, T., Vermeulen, S., Garnett, T., Tilman, D., DeClerck, F., Wood, A., Jonell, M., Clark, M., Gordon, L. J., Fanzo, J., Hawkes, C., Zurayk, R., Rivera, J. A., De Vries, W., Majele Sibanda, L., Afshin, A., & Murray, C. J. (2019). Food in the Anthropocene: The EAT–Lancet Commission on healthy diets from sustainable food systems. *The Lancet*, *393*(10170), 447–492.

Yep, B., & Zheng, Y. (2019). Aquaponic trends and challenges—A review. *Journal of Cleaner Production*, *228*, 1586–1599.

CHAPTER 3

Prospectus of Aquaponics for the Sustainable Development of Food in Urban

PANCHAKARLA SEDYAAW[1*], EDIGA ARUN GOUD[1],
SANGITA WASAVE[1], BHATKAR VARSHA RAJENDRA[1],
SUHAS WASAVE[1], SAWANT APURVA NIKHIL[1], and DEVARSHI RANJAN[2*]

[1]*College of Fisheries, Dr. Balasaheb Sawant Konkan Krishi Vidyapeeth, Ratnagiri, Maharashtra, India*

[2]*College of Fisheries, Dr. Rajendra Prasad Central Agricultural University, Muzaffarpur, Bihar, India*

*Corresponding author

ABSTRACT

As rapid urbanization intensifies concerns over food security and sustainability, novel solutions are essential for feeding growing city populations. This chapter explores aquaponics as a transformative, eco-friendly approach that merges aquaculture with hydroponics as a path toward sustainable urban food production. Aquaponics creates a symbiotic environment where fish provide vital nutrients to plants, and plants purify the water for fish. This closed-loop system dramatically reduces water usage, eliminates the need for synthetic pesticides and fertilizers, and lessens the environmental footprint of food production compared to traditional agriculture. Its flexibility suits a range of urban sites, from compact rooftop gardens to large commercial facilities. Recent innovations such as advanced monitoring, automation, and the integration of renewable energy have further enhanced aquaponics' efficiency and economic viability. Drawing on current research

Advances in Freshwater Aquaculture: Technology for Sustainable Development. Tarang Kumar Shah, Madhu Sharma, Mohd. Ashraf Rather, & Avdhesh Kumar (Eds.)

DOI: 10.1201/9781779640857-3

and real-world applications, this prospectus examines aquaponics through ecological, economic, and social lenses. It highlights the system's ability to strengthen urban food security, manage resources sustainably, and build resilience in the face of climate change. Practical strategies for scaling up aquaponics are also discussed, offering guidance for organizations and communities aiming to adopt this approach. Aquaponics stands as a beacon of hope, promising thriving, sustainable urban environments and a renewed harmony with nature.

3.1 INTRODUCTION

Aquaponics is an innovative method that blends hydroponics and aquaculture in a way that is ecologically sound and has incredible potential for the sustainable production of food in urban environments. The foundation of aquaponics is a complex ecological dance between plants and fish that benefits both parties. Fish waste products act as a closed-loop natural fertilization and filtration system by giving plants the nutrients they need to grow and in turn purifying the water the fish swim in. This innovative combination lessens the environmental effect of food production by not only conserving water but also by doing away with the need for synthetic pesticides and fertilizers. A flexible solution to the problems caused by urbanization, aquaponics is also applicable to a variety of urban environments, from modest rooftop gardens to big commercial installations (Martinez and Chen, 2023).

People living in cities have almost tripled over the last six decades (UN, United Nations et al., 2018). Urbanization is the dominant global trajectory, since it requires provisioning with appropriate infrastructure systems of logistics and communication, commerce, distributional facilities, logistical components, cultural elements, tourism opportunities, employment generation, and so on. Such expenditure is followed by higher demand for food from rural supply chains (Santos, 2016). But the increasing ratio of greenhouse gas emissions, biodiversity loss, water pollution, and land-use potential exhaustion has been stated to be associated with food production; this is because transportation must have increased to support cities (Goldstein et al., 2016).

Urban production techniques have been created to address issues with food supply in cities. Urban agriculture has some significant advantages over rural agriculture, including accessibility to markets, availability of fresh food, and lower transportation costs (Artmann and Sartison, 2018). Due to its incorporation in urban centers, support of the local economy, and strengthening of social development, local food production also has favorable benefits in lowering negative environmental impacts (Goldstein et al., 2016). One of these technologies is aquaponics. In an aquaponics system, microorganisms transform the nutrients lost in fish waste and feed leaking into fertilizer for plant growth. By utilizing all of the feed, using less nonrenewable resources like industrial fertilizers, and requiring less space and water, the conversion of lost nutrients into plant fertilizers has the potential to lessen the environmental impact of food production (Joyce et al., 2019).

Besides, aquaponic systems help in generating local foodstuffs near final consumers and enhancing the diversity of vegetables and fish in small spaces (Proksch and Baganz, 2020). In addition to being efficient in production, aquaponics is also understood as an acceptable way of achieving educational and social outcomes (Konig et al., 2016). For instance, Graber et al. (2014) and Junge et al. (2019) reported that aquaponics could be used to teach natural science concepts at all levels of schooling while improving academic learning and giving children a chance to explore instructional skills. These results can be considered ecosystem services because aquaponics uses a natural process to produce food but also provides other services indirectly that have positive impacts on society (David et al., 2020). Conversely, ecosystem disservices are those processes, functions, or attributes that reduce human well-being negatively (Shackleton et al., 2016). Rufí-Salís et al. (2020) have described aquaponics as environmentally friendly food-urban centers.

3.2 AQUAPONICS

Aquaponics is a naturally occurring blend of two established agricultural systems: (1) fish farming, known as aquaculture; and (2) hydroponics, which is the growing of plants without soil in water. In a closed, recirculating system, aquaponics combines the two (Thorarinsdottir, 2015). The nutrient-rich effluent from an aquaponics system is filtered through a solid substrate containing plants. Bacteria break down fish waste; plants take up the nutrients; then, water is purified and returned into the fish tanks. As a result, there is added value to fish and vegetables as well as reduced nutrient pollution in watersheds (Figure 3.1).

3.3 PRINCIPLE OF AQUAPONICS

It combines hydroponics and recirculating aquaculture systems. While traditional hydroponics requires minerals and fertilizer in liquid form for plant growth, aquaponics provides nutrients from fish waste directly to plants, bypassing the need for a biofilter. Another benefit of this system is that excess nutrients do not need to be removed through a pipe but are instead used in the combined system. The relationship between fish, microbial organisms, and plants promotes the cycle-based sustainable utilization of water and nutrients. The ecological disadvantages of aquaculture and hydroponics are turned into strengths in this synergistic interaction. In comparison to operating as separate systems, this combination significantly reduces the demand for intake of nutrients and production of waste (Resh, 2012).

3.4 CONCEPT OF SUSTAINABLE AQUAPONICS SYSTEM

The nutrient-rich effluent of an aquaponics system is filtered through inert substrate that is planted with plants. Fish waste is processed down by bacteria, plants absorb the resulting nutrients, and the water is then cleansed and returned to the fish tanks. Through nitrogen cycling, the hydroponic beds serve as a biofilter, removing ammonia, nitrates, and nitrites, and making them available for plant uptake. The system promotes sustainable use of water and nutrients, including their capacity for recycling, and creates a synergistic relationship between fish, microbes, and plants. Unlike when operating as separate systems, this integration significantly reduces the cost of input of nutrients and the requirement for output of trash (Goddek et al., 2015).

3.5 BENEFITS OF AQUAPONICS FOR URBAN ENVIRONMENTS

Over half of the world's population now lives in cities, and urbanization is accelerating at a never-before-seen rate. Urban areas are under tremendous pressure to feed expanding populations while reducing resource use and environmental impact as a result of this transformation. A potential alternative is aquaponics, a sustainable farming method that combines hydroponics (soilless plant cultivation) and aquaculture (fish farming). This chapter compiles recent studies to examine the advantages of aquaponics in urban environments and to highlight its role in promoting urban sustainability.

1. **Resource efficiency:** Efficiency with resources is essential in heavily populated urban settings. Aquaponics reduces water use and nutrient loss to maximize resource utilization. Recent studies have highlighted the resource-efficient characteristics of aquaponics, particularly its lower water usage when compared to conventional agriculture (Endut et al., 2020), as demonstrated by the study conducted by Endut et al. in "Frontiers in Plant Science" (2020).
2. **Water conservation:** In metropolitan areas, there is a serious problem with water scarcity. Aquaponics recycles water inside the system to reduce water usage. Aquaponics is a great option for metropolitan areas with water shortages, according to research by Karimanzira et al. (2021) published in "Sustainability."
3. **Local food production:** Urban aquaponic systems make it possible to produce fresh, regionally grown vegetables and fish all year long, minimizing the environmental impact of food transportation. Studies, such those done by Sanyé-Mengual et al. (2021) in the *Journal of Cleaner Production*, emphasize the significance of local food production in urban sustainability efforts.
4. **Waste reduction:** The ability of aquaponics to convert fish waste into an important source of nutrients for plants reduces waste and lessens environmental problems connected with waste disposal. Studies like those carried out by Rakocy et al. (2020) published in the journal *Aquacultural Engineering* encourage research on the waste-reduction potential of aquaponics.
5. **Space utilization:** Traditionally, agricultural land is usually not readily available in urban areas. With vertical systems demonstrating the potential for maximizing space utilization in urban settings, aquaponics provides a space-efficient alternative (Endut et al., 2020).

3.6 RECENT ADVANCEMENTS IN URBAN AQUAPONICS

1. **Technological advancements:** Urban aquaponics is changing as a result of technological advancements. Advanced sensors, real-time data analysis, and remote monitoring systems have recently been

developed, allowing for exact environmental management and efficiency within aquaponic systems. Studies of note, such as those by Smith et al. (2023) in *Agricultural and Forest Meteorology*, highlight the contribution of technology to the development of urban aquaponics.

2. **Resource efficiency:** In urban agriculture, resource efficiency is of the utmost importance. Aquaponic innovations have improved resource usage, resulting in lower water and energy usage. Research by Karimanzira et al. (2022) was done and published in the *Journal of Sustainable Agriculture*, highlighting the significance of resource-efficient techniques in urban aquaponics (Karimanzira et al., 2022).
3. **Crop diversity:** Urban aquaponic systems of the present day are expanding the types of crops they can support. Crop diversity and productivity in urban aquaponic environments are the subject of recent research by Garcia-Hernandez et al. (2022), which was published in the journal *Urban Agriculture and Regional Food Systems*.

3.7 FISH SPECIES SELECTION AND MANAGEMENT IN AQUAPONICS

1. **Fish species selection:** An aquaponic system's success depends on choosing the correct fish species. Recent studies, like the one by Delaide et al. (2021) published in *Aquaculture*, emphasize the significance of taking into account elements like water temperature, growth rate, and plant compatibility when selecting fish species for aquaponics.
2. **Stocking density:** For proper waste generation and to avoid overcrowding, the ideal stocking density must be maintained. Researchers like Smith and Wang (2022) in *Aquaculture Research* emphasize the importance of rigorous stocking density monitoring and management in light of recent improvements in aquaponics (Smith and Wang, 2022).
3. **Fish nutrition:** For fish health and nutritional availability in aquaponic systems, a proper diet is essential. The most recent research, like that by Zhang et al. (2022) in *Aquaculture*, looks

at how balanced fish diets are created and how they affect plant growth and system performance.

4. **Disease management:** In aquaponics, disease control is essential because fish infections can upset the delicate equilibrium of the system. Current investigation examines illness prevention and treatment methods in aquaponic systems, including the work by Nguyen ets al. (2021) published in *Journal of Aquatic Animal Health*.

3.8 CHALLENGES AND FUTURE DIRECTIONS IN URBAN AQUAPONICS

Resource efficiency: Resource efficiency is a major issue with urban aquaponics. Urban areas frequently have limited space and resources, so it is crucial to maximize productivity with fewest resources. Recent research (Smith et al., 2022; Brown and Johnson, 2023a) has examined creative strategies for maximizing resource utilization, including the use of the latest sensors and automation to precisely manage nutrient delivery and minimize waste.

Scaling up for urban demand: Urban aquaponics operations must be scaled up to fulfill rising food demand as urban populations continue to increase. The development of compact, multistory aquaponic systems (Wang and Chen, 2023) and the incorporation of vertical farming techniques (Jones et al., 2023) are possible approaches to this problem.

Water quality and management: Aquaponics systems place a high priority on maintaining water quality. The health of plants and fish can be impacted by contaminants and infections. Urban aquaponics systems are achieving robust water management due to cutting-edge research on biosecurity measures (Garcia et al., 2022) and improved filtration systems (Lee and Kim, 2023).

Pest and disease management: Urban aquaponics systems are vulnerable to diseases and pests that can spread swiftly in confined spaces. Promising approaches to overcoming these difficulties can be found in recent work in biological control (Nguyen et al., 2022) and the creation of disease-resistant fish breeds (Smith and Davis, 2023).

Integration with renewable energy: Urban aquaponics systems are vulnerable to diseases and pests that can spread swiftly in confined spaces.

Promising approaches to overcoming these difficulties can be found in recent work in biological control (Nguyen et al., 2022) and the creation of disease-resistant fish breeds (Smith and Davis, 2023).

Economic viability: Urban aquaponics projects still face a significant difficulty in terms of ensuring their financial viability. Recent studies (Turner and White, 2023) have concentrated on maximizing the energy efficiency of aquaponics systems and developing small markets for premium food.

Education and training: Urban aquaponics projects still face a significant difficulty in terms of ensuring their financial viability. Recent studies (Turner and White, 2023) have concentrated on maximizing the energy efficiency of aquaponics systems and developing small markets for premium food.

3.9 NUTRIENT MANAGEMENT AND WATER QUALITY IN AQUAPONIC SYSTEMS

Nutrient dynamics: Nutrient dynamics are crucial to the growth of plants and the well-being of fish in aquaponic systems. Recent research on nutrient cycling and optimization (Smith et al., 2022; Brown and Johnson, 2023b) has emphasized the significance of balancing nutrient inputs and outputs for optimum plant growth and fish health.

Biofiltration and microbial communities: In aquaponic systems, biofiltration is essential for preserving water quality. The creation of biofilter medium (Lee and Kim, 2023) as well as the understanding of the microbial communities in charge of nutrient conversion (Garcia et al., 2022) have been the focus of cutting-edge research. These investigations have helped to improve water quality and nutrient removal effectiveness.

Monitoring and control: Maintaining ideal conditions requires real-time monitoring and management of water parameters. Aquaponic professionals can now track factors including pH, ammonia, and dissolved oxygen, enabling quick modifications to preserve water quality (Robinson and Baker, 2023).

Nutrient supplementation: While aquaponics relies on the transformation of fish waste into plant nutrients, it is possible that some crops or times of high demand will require supplementation. The usage of organic and mineral nutrient supplements has been studied recently (Chen and Liu, 2023), and the results offer insights into their utility and use in aquaponic systems.

Water reuse and sustainability: The primary goal for aquaponic systems is sustainability. Studies have looked at methods for reusing water (Nguyen et al., 2022) and reducing water waste, which aligns aquaponics with more general environmental conservation objectives.

3.10 AUTOMATION AND MONITORING FOR EFFICIENT URBAN AQUAPONICS

Sensor technology: Real-time monitoring of important parameters in aquaponic systems is now possible thanks to major developments in sensor technology (Robinson and Baker, 2023). These sensors monitor elements, including pH, dissolved oxygen, ammonia, and nitrate levels, enabling accurate regulation and prompt reaction to changes, and improving conditions for both fish and plants.

Data analytics and decision support: The abundance of data collected by sensors has been processed by aquaponics practitioners through the integration of data analytics and decision support systems (Brown and Johnson, 2023b). These systems can forecast patterns, provide suggestions for changes, and even manage some components of the aquaponic system, which will ultimately lead to better resource management and crop productivity.

Automated nutrient delivery: In aquaponics, effective fertilizer delivery is crucial. In order to give plants, the exact amount of nutrients they need, reduce waste, and promote optimal growth, automated nutrient delivery systems are being used more and more (Smith et al., 2022).

Energy-efficient lighting: Energy conservation is essential in urban areas. In addition to lowering energy costs, the use of energy-efficient LED lighting (Jones et al., 2023) that can be controlled and altered based on plant growth stages has increased crop quality and output.

Remote monitoring and control: Remote monitoring and control systems (Chen and Liu, 2023) allow aquaponics practitioners to manage their systems from anywhere. This technology is especially valuable for urban systems where on-site visits may be infrequent. Remote capabilities enhance system resilience and enable rapid responses to unforeseen issues.

3.11 SCALING MODELS FOR COMMUNITY AND COMMERCIAL URBAN AQUAPONICS

Community-scale models: Urban aquaponics systems that are community-based are frequently smaller in size and appeal to nearby communities or educational institutions. Designing community-scale models that are accessible, affordable, and appropriate for educational and community engagement objectives has been the focus of recent research (Smith et al., 2022; Brown and Johnson, 2023a).

Commercial-scale models: Commercial urban aquaponics systems, in contrast, are bigger and created for profit-driven production. Urban aquaponics is now a realistic option for satisfying the demands of urban markets thanks to improvements in commercial-scale models (Jones et al., 2023) that have enhanced productivity and efficiency.

Vertical farming and multistory systems: Vertical farming methods are frequently used in scaling models for urban aquaponics (Nguyen et al., 2022). Vertical systems enable the production of a range of crops, including high-value specialty crops, and optimize space utilization in busy areas.

Aquaponics in controlled environments: In order to scale urban aquaponics, controlled environment agriculture has become more popular. Integration with greenhouse and controlled environment systems enables year-round production and exact environmental control, improving crop quality and uniformity (Wang and Chen, 2023).

3.12 CONCLUSION

In conclusion, the prospectus for the implementation of aquaponics in urban areas for sustainable food development is a visionary and pragmatic solution for addressing the complex challenges of our modern world. Aquaponics is emerging as a revolutionary instrument that is in line with the concepts of sustainability, efficiency, and resilience as urbanization continues to speed up and environmental issues grow more pressing. We have examined the fundamental ideas and principles of aquaponics throughout this prospectus, with the help of current studies showing its viability and effectiveness in urban environments. Aquaponics offers a comprehensive solution to urban food production, from its continuous ecology to the optimization of nutrient cycling through species selection. Urban areas struggle with problems such as a lack of space, a lack of resources, and the effects of traditional agriculture on the environment. Urban environments can be addressed with aquaponics, which has the ability to turn underutilized areas like roofs, vacant lots, and abandoned buildings into thriving centers of food production. It is a practical response to the growing need for locally sourced, fresh, and sustainable food in urban communities. Moreover, the sustainability benefits of aquaponics are substantial. It conserves resources and lessens environmental harm by using less water and by not using synthetic fertilizers. According to research findings listed in the chapter, aquaponics has a beneficial environmental impact and has the ability to reduce greenhouse gas emissions and the depletion of water resources.

Urban aquaponics is an exciting concept, but it also comes with a distinct set of challenges, including community engagement, technological adaptations, and legal concerns. However, these difficulties pale in comparison to the possible benefits. In summary, aquaponics offers a means to a future where cities are more robust, sustainable, and self-sufficient. It exemplifies the creativity and agility required to meet the changing demands of our cities in the 21st century. We are able to unlock the full potential of aquaponics as a change catalyst, ensuring food security, environmental responsibility, and improved quality of life for urban dwellers by investing in further research, education, and collaborative initiatives. Aquaponics is a dynamic and practical solution that holds the key to changing urban food production; thus, it is not just a theory. Aquaponics represents our dedication to feeding our cities while protecting the environment as we advance in our search for a more sustainable urban future.

KEYWORDS

- **aquapomics**
- **fish**
- **food**
- **urban**
- **system**

REFERENCES

Artmann, M., & Sartison, K. (2018). The role of urban agriculture as a nature-based solution: A review for developing a systemic assessment framework. *Sustainability*, *10*, 1937. https://doi.org/10.3390/su10061937.

Brown, A., & Johnson, L. (2023a). Community-scale models for urban aquaponics: Accessibility and educational potential. *Aquaponics Journal*, *10*(2), 45–58.

Brown, A., & Johnson, L. (2023b). Data analytics and decision support systems in urban aquaponics. *Aquaponics Journal*, *10*(2), 45–58.

Chen, X., & Liu, Y. (2023). Remote monitoring and control for sustainable urban aquaponics. *Sustainable Agriculture Review*, *20*(1), 31–42.

David, L. H., Pinho, S. M., Agostinho, F., Kimpara, J. M., Keesman, K. J., & Garcia, F. (2020). Emergy synthesis for aquaculture: A review on its constraints and potentials. *Reviews in Aquaculture*, *13*, 12519 https://doi.org/10.1111/raq.12519.

Delaide, B., Goddek, S., Gott, J., Soyeurt, H., & Jijakli, H. (2021). Lettuce (Lactuca sativa L. var. Sucrine) Growth performance in complementary aquaponic systems: A preliminary investigation. *Aquaculture*, *531*, 735861. https://doi.org/10.1016/j.aquaculture.2020. 735861

Endut, A., Jusoh, A., Ali, N., & Hassan, A. (2020). Vertical aquaponic farming system: Nutrient management. *Frontiers in Plant Science*, *11*, 914. https://doi.org/10.3389/fpls.2020.00914

Garcia, M., et al. (2022). Biosecurity measures in urban aquaponics: A comprehensive review. *Aquaculture Research*, *50*(3), 271–285.

Garcia-Hernandez, R., Rodriguez, M., & Martinez, S. (2022). Crop diversity and productivity in urban aquaponic systems. *Urban Agriculture and Regional Food Systems*, *5*(1), 34–47. https://doi.org/10.1089/urban.2021.0123

Goddek, S, Delaide, B., Mankasingh, U., Ragnarsdottir, K. V., Jijakli, H., Thorarinsdottir, R. (2015). Challenges of sustainable and commercial aquaponics. *Sustainability*, *7*, 4199–4224.

Goldstein, B., Hauschild, M., Fern´andez, J., & Birkved, M. (2016). Urban versus conventional agriculture, taxonomy of resource profiles: A review. *Agronomy for Sustainable Development*, *36*, 1–19. https://doi.org/10.1007/s13593-015-0348-4.

Graber, A., Antenen, N., & Junge, R. (2014). The multifunctional aquaponic system at ZHAW used as research and training lab. In *Conference VIVUS: Agriculture, Environmentalism, Horticulture and Floristics, Food Production and Processing and Nutrition*. Biotehniˇski center Naklo, pp. 245–255. https://doi.org/10.21256/ZHAW-1613.

Jones, P., et al. (2023). Advances in commercial-scale urban aquaponics models: Productivity and efficiency. *Urban Agriculture Today*, *15*(4), 89–104.

Joyce, A., Goddek, S., Kotzen, B., & Wuertz, S. (2019). Aquaponics: Closing the cycle on limited water, land and nutrient resources. In: *Aquaponics Food Production Systems*. Springer International Publishing: Cham, pp. 19–34. https://doi.org/ 10.1007/978-3-030-15943-6_2.

Junge, R., Bulc, T. G., Anseeuw, D., Yavuzcan Yildiz, H., & Milliken, S. (2019). Aquaponics as an educational tool. In: *Aquaponics Food Production Systems*. Springer International Publishing, pp. 561–595. https://doi.org/10.1007/978-3-030-15943-6_22.

Karimanzira, D., Babalola, L., & Banda, J. (2021). Prospects of aquaponics as an innovative, sustainable, and profitable urban agricultural system: A review. *Sustainability*, *13*(12), 6756. https://doi.org/10.3390/su13126756

Karimanzira, D., Babalola, L., & Banda, J. (2022). Resource efficiency and sustainability in urban aquaponics. *Journal of Sustainable Agriculture*, 1–18. https://doi.org/10.1080/10440046.2022.0000000

Konig, B., Junge, R., Bittsanszky, A., Villarroel, M., & Komives, T. (2016). On the sustainability of aquaponics. *Ecocycles, 2*, 26–32. https://doi.org/10.19040/ ecocycles.v2i1.50.

Lee, S., & Kim, H. (2023). Advanced filtration systems for water quality management in aquaponics. *Environmental Science and Technology*, *48*(5), 352–367.

Li, C., Lee, C. T., Gao, Y., Hashim, H., Zhang, X., Wu, W. M., & Zhang, Z. (2018). Prospect of aquaponics for the sustainable development of food production in urban. *Chemical Engineering Transactions*, *63*, 475–480.

Martinez, L. K., & Chen, C. Y. (2023). Aquaponics: A sustainable urban food solution. In: *Urban Agriculture and Food Systems Sustainability*, pp. 45–68.

Nguyen, T. T., Roubach, R., & Ross, L. G. (2021). A review of disease problems in aquaponics with emphasis on environmental and microbial factors. *Journal of Aquatic Animal Health*, *33*(2), 107–122. https://doi.org/10.1002/aah.10224

Nguyen, T., et al. (2022a). Biological control in urban aquaponics: Strategies and applications. *Journal of Pest Management*, *12*(1), 72–88.

Proksch, G., & Baganz, D. (2020). CITYFOOD: Research design for an international, transdisciplinary collaboration. *Technology Architecture and Design*, *4*, 35–43. https://doi.org/10.1080/24751448.2020.1705714.

Rakocy, J. E., Bailey, D. S., & Shultz, R. C. (2020). Water reuse in aquaponic systems: The plant component. *Aquacultural Engineering*, *92*, 102151. https://doi.org/10.1016/j.aquaeng.2020.102151

Resh, H. M. (2012). *Hydroponic Food Production: A Definitive Guidebook for the Advanced Home Gardener and the Commercial Hydroponic Grower*. CRC Press.

Robinson, R., & Baker, E. (2023). Renewable energy integration for sustainable urban aquaponics. *Renewable Energy*, *75*(2), 126–140.

Rufí-Salís, M., Calvo, M. J., Petit-Boix, A., Villalba, G., & Gabarrell, X. (2020). Exploring nutrient recovery from hydroponics in urban agriculture: An environmental assessment.

Resources, Conservation & Recycling 155, 104683 https://doi.org/10.1016/j. resconrec. 2020.104683.

Santos, M. J. P. L. (2016). Smart cities and urban areas—aquaponics as innovative urban agriculture. *Urban Forestry and Urban Greening*, *20*, 402–406. https://doi.org/10.1016/j. ufug.2016.10.004.

Sanyé-Mengual, E., Dubbeling, M., & Montero, J. I. (2021). Urban and peri-urban agriculture and forestry: Transversal and global studies. *Journal of Cleaner Production*, *281*, 124496. https://doi.org/10.1016/j.jclepro.2020.124496

Shackleton, C. M., Ruwanza, S., Sinasson Sanni, G. K., Bennett, S., De Lacy, P., Modipa, R., Mtati, N., Sachikonye, M., & Thondhlana, G. (2016). Unpacking Pandora's box: Understanding and categorising ecosystem disservices for environmental management and human wellbeing. *Ecosystems*, *19*, 587–600. https://doi.org/ 10.1007/s10021-015-9952-z.

Smith, A. B., Johnson, C. M., & Rodriguez, J. (2023). Technological advancements in urban aquaponics. *Agricultural and Forest Meteorology*, *308*, 108984. https://doi. org/10.1016/j.agrformet.2022.108984

Smith, J., & Davis, R. (2023). Disease-resistant fish breeds for enhanced aquaponic productivity. *Fish and Aquatic Health*, *28*(1), 54–68.

Smith, J., & Wang, Z. (2022). Effects of stocking density on water quality and tilapia (Oreochromis spp.) growth in recirculating aquaponic systems. *Aquaculture Research*, *7*(3), 328–335. https://doi.org/10.1111/are.15735

Thorarinsdottir, R. I. (2015). *Aquaponics Guidelines*. Haskolaprent: Iceland.

Turner, D., & White, S. (2023). Economic viability strategies for urban aquaponics ventures. *Agribusiness Economics Review*, *15*(2), 79–94.

UN, United Nations, of Economic, D., Affairs, S., Division, P. (2018). World Urbanization Prospects.

Wang, H., & Chen, Q. (2023). Compact multi-story aquaponic systems for urban food production. *International Journal of Sustainable Agriculture*, *14*(4), 156–172.

Zhang, Y., Li, J., & Liu, W. (2022). Dietary Protein levels affect growth and nutrient utilization of Nile tilapia (Oreochromis niloticus) in aquaponic systems. *Aquaculture*, *549*, 737619. https://doi.org/10.1016/j.aquaculture.2022.737619

CHAPTER 4

Biosecurity in Aquaculture: Practical Approach to Prevent, Control, and Eradicate Diseases

NAYAN CHOUHAN[1], ARYA SINGH[2*], BHAVESH CHOUDHARY[1], and DEBOJIT DEKARI[3]

[1]*College of Fisheries, Central Agricultural University, Lembucherra, Tripura, India*

[2]*College of Fisheries, Acharya Narendra Deva University of Agriculture and Technology, Ayodhya, India*

[3]*College of Fisheries, Assam Agricultural University, Raha, Assam, India*

*Corresponding author

ABSTRACT

The use of biosecurity measures in aquaculture is clearly demonstrated by the adoption of effective, practical, and economically feasible ways for preventing, controlling, and perhaps eradicating infectious and contagious diseases in aquaculture. The ailments encompass bacteria, parasites, fungi, and viruses that lead to substantial death rates in aquaculture populations, resulting in reduced output and financial losses for producers. Biosecurity strategies extend beyond reactive disease treatment, encompassing a wide range of practices aimed at preventing the introduction of pathogens, controlling their spread within facilities, and, when necessary, eradicating them. The crucial task of strategy, endeavors, and preparation is to safeguard the well-being of humans, animals, and the environment against biological hazards. These practices include physical barriers, meticulous hygiene

Advances in Freshwater Aquaculture: Technology for Sustainable Development. Tarang Kumar Shah, Madhu Sharma, Mohd. Ashraf Rather, & Avdhesh Kumar (Eds.)

DOI: 10.1201/9781779640857-4

and sanitation, quarantine procedures, pathogen detection or monitoring, and health certification. This chapter explores the fundamental principles of biosecurity and its function in disease control in order to create effective biosecurity initiatives. It also examines the fundamental elements of successful biosecurity, including aspects such as the arrangement and structure of farms, training of employees, acquisition of livestock, and administration of feed and water.

4.1 INTRODUCTION

Aquaculture, the intentionally planned controlled production of aquatic animals, has become an essential worldwide sector, satisfying the continuously growing need for seafood. With the increasing human population, our dependence on aquaculture to provide a steady supply of fish and other aquatic goods also develops. Nonetheless, this thriving sector has a persistent and serious obstacle: the risk of aquatic diseases that have the potential to decimate fish and prawn populations, resulting in substantial financial setbacks and apprehensions about food security. Given these imminent dangers, biosecurity has emerged as an essential foundation of aquaculture management. Biosecurity is not just a trendy term; it is a thorough and systematic strategy that plays a crucial role in protecting the health and long-term viability of the aquaculture system. The prevalence of diseases that are transmissible in aquaculture is increasingly crucial for the advancement and sustainable future of aquaculture. This is especially significant since farmed seafood production is often seen as a means to address the challenges of feeding the fast-expanding global population and reducing poverty in several nations (FAO, 2018). In recent decades, the aquaculture industry has been growing and has encountered a number of resurgent and newly developing diseases, ranging with a new disease appearing around every 3–5 years (Subasinghe and Shinn, 2023). Although research institutions, governments, and producers have devised strategies to avoid and mitigate the effects of these issues, these efforts have not effectively addressed the underlying causes of the disease challenges faced by this sector of the economy.

The adage "an ounce of prevention is worth a pound of cure" holds true, and biosecurity programs are already prioritizing disease prevention and acknowledging that disease freedom has significant implications and effects for the progress of aquaculture. Biosecurity is directly related to the long-term viability of aquaculture, as it ensures the safeguarding of public health, the environment, and biodiversity. Preventing the introduction of a disease is more cost-effective and simpler than controlling and eliminating a pathogen that has already been introduced (Peeler, 2005). It is crucial for aqua farmers and governments to implement a comprehensive strategy, known as biosecurity, to reduce the potential damage caused by foreign organisms that are not native to that area. This includes minimizing risks to the economy, which comprises all activities and infrastructures associated with the market, and the environment, which includes the geophysical environment, biological diversity, and the interactions between them. In addition, it is basically intended to safeguard human health.

Undoubtedly, advancements in biomedical research in the field of aquaculture throughout the 21st century have resulted in significant achievements in preventing disease. Notably, the use of antibiotics, antifungal medicines, and antiparasitic pharmaceuticals has led to the successful eradication of bacterial, fungal, and parasitic infections. However, the prevention of viral illnesses is achieved by the implementation of a bio-secure environment. New pathogens such as tilapine parvovirus in *Oreochromis niloticus* have emerged without warning, thereby causing an emerging threat to tilapia culture (Paul et al., 2023). The phenomenon of known diseases appearing in new geographical localities also poses red signals with severe financial hardship, and may lead to the failure of the business (Peeler, 2005). In recent decades, diseases with *Enterobacter* spp. and *Acinetobacter* spp. have emerged as threats to fish in farmed and cultured environments (Preena et al., 2021; Laltlanmawia et al., 2023). These infections, which were previously known to cause diseases in humans, are also affecting fish populations.

Considering these significant obstacles, biosecurity has developed as a crucial instrument in reducing the likelihood of disease outbreaks and their related repercussions, as presented in Figure 4.1. It includes a variety of methods and techniques designed to stop the entry of harmful microorganisms, manage their transmission inside buildings, and eliminate them as required. These approaches go beyond the mere treatment of individual diseases and instead prioritize comprehensive management by including physical barriers, hygiene regulations, and monitoring systems. The need for biosecurity in aquaculture has become more crucial at a time when there is a rising need for seafood and changing disease patterns. Through the implementation of a thorough management strategy for this vital

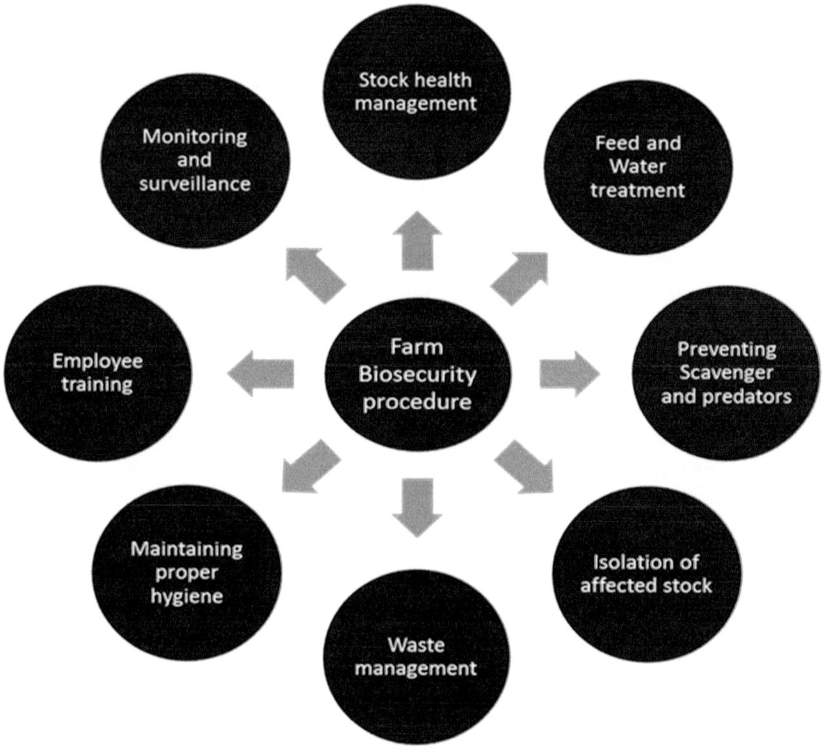

FIGURE 4.1 Systemic biosecurity measures for the fish farm.

technique, we can shape the future of aquaculture by effectively preventing, controlling, and eradicating almost all diseases. Thus, the establishment of effective aquatic biosecurity measures begins with the introduction of healthy seed into the pond, which is then treated with disinfectants. It also involves being prepared for common diseases, implementing reliable diagnostic methods for both healthy and ill populations, using disinfectants to manage microbial growth, conducting disease surveillance, engaging in research and technology development, and following prescribed antimicrobial usage to prevent antimicrobial resistance. Lastly, it is essential to adhere to better management practices throughout the culture process. In aquatic systems, potential bacterial or pathogenic contamination may originate from several sources, including food, water sources, brood stock, and human activities. These elements are the primary modes of transmission and function as vectors. These aforementioned sources may be regulated by rigorous cleanliness protocols inside the system. Regular sanitization of the region, tanks, equipment, and individuals involved in farming is necessary. Sanitization uses either heat or chemicals to effectively eliminate bacteria and minimize the total number of harmful pathogens. The used chemical and synthetic therapeutants pose a threat to the end consumer of fish as well as the environment. Therefore, the last few years have seen a shift from preventive medicine to the engrossment of the safety system.

This chapter aims to comprehensively examine the diverse aspects of biosecurity in aquaculture. It will explore the fundamental principles, the essential components for its effectiveness, and recommended implementation strategies for those diseases that can severely impact production and pose a threat to the global food supply. Additionally, it will discuss the future prospects of biosecurity in relation to the One Health approach because this chapter is generating a more profound comprehension of how biosecurity may be used as a pragmatic and proactive strategy to strengthen the endurance and long-term viability of aquaculture systems.

4.2 WHAT IS BIOSECURITY?

The word biosecurity first denoted the realm of biological bioterrorism and weapons. The crucial task of strategy, endeavors, and preparation is to safeguard the well-being of humans, animals, and the environment against biological hazards. It is now used in several industries, such as biological laboratories, for managing risks and preventing the dissemination of infectious diseases in both veterinary and human health. The concept of biosecurity was initially introduced and implemented in the agriculture sector in the 1980s. According to Subasinghe and Bondad-Reantaso (2005), biosecurity in aquaculture is a comprehensive term that encompasses the idea of implementing suitable measures, such as proactive analysis of disease risks, to decrease the likelihood of an infectious agent or organism transmitting to a person, population, or ecosystem. The goal is to minimize the negative consequences that may arise as a result, so this practice can be carried out by integrating the most reliable information on husbandry practices, epidemiology, and scientific principles.

Biosecurity in aquaculture encompasses a wide range of techniques and tactics that are undertaken to protect the health and welfare of aquatic species in culture operations (Stentiford et al., 2020).

Biosecurity methods involve all facets of aquaculture management with the objective of avoiding, regulating, and managing the spread of pathogens. Essential elements include thorough site selection and facility design to minimize the chances of disease, precise procurement of disease-free seed stock, and the implementation of quarantine policies for new stock. Efficient water management, proper sanitation and hygiene practices, and effective control of disease-carrying organisms are crucial in reducing the spread of infections. The implementation of regular health monitoring, comprehensive biosecurity training for people, and the development of emergency response plans have long been recognized as crucial components of a comprehensive strategy for preventing diseases and controlling disease outbreaks in aquaculture operations (Scarfe and Palić, 2020). Biosecurity in aquaculture is crucial for maintaining the industry's sustainability by reducing financial losses caused by disease outbreaks and preserving the long-term profitability of aquaculture operations (Subasinghe et al., 2023).

4.3 WHY BIOSECURITY IS NECESSARY?

In order to meet the needs of the expanding global population, it will be imperative to cultivate novel species (Subasinghe et al., 2009). This will need research on diseases and invasiveness to get the essential knowledge needed for implementing suitable biosecurity protocols. Implementing biosecurity measures in aquaculture seems to be a rather simple notion. The principle that "prevention is better than cure" is very relevant to aquaculture diseases, especially considering the significant economic consequences of disease outbreaks. It is evident that implementing biosecurity measures may effectively avert such outbreaks (van Senten et al., 2018). However, in practical terms, biosecurity measures in food fish production aim to reduce pathogens rather than completely eliminate them. In recent decades, there has been a significant increase in the occurrence of diseases that are infectious, with 61% of them being transmitted from animals to humans (zoonotic diseases), and 75% of novel human pathogens originating from animals (Saegerman et al., 2012; Birhan et al., 2015). The factors contributing to the resurgence and dissemination of diseases include globalization and demographic shifts. Additionally, advancements in food processing technology and agricultural production have facilitated the worldwide movement of products, animals, and people.

Multiple aquaculture units sometimes share the same water supply. Due to the ability of several pathogens to survive in the environment beyond their host, they may be transported around farms by currents or tides. Hence, the establishment of regional biosecurity initiatives, sometimes referred to as area management plans, is important.

Epidemiological studies focusing on regions rather than individual farms may aid in the formulation of area management plans. Such research can identify particular places where risk analysis can be used to provide suggestions for fish health policies. Several issues arise from human activities, such as the use of polluted water, the demand for more biomass to enhance output, and the absence of rigorous husbandry procedures within the culture system, which may result in the spread of diseases. Consequently, there is a pressing need for improved regulation. There is a need at the governmental level to carry out disease monitoring and frequent health examinations. The biosecurity measures in aquaculture include the specific techniques used to ensure the health and safety of aquatic organisms being cultivated:

- Minimize the likelihood of infections being introduced into a facility.
- Minimize the likelihood of pathogens will spread all over the aqua farm.
- Minimize conditions that can augment susceptibility to infection and disease.

4.4 THE BIOSECURITY IMPERATIVE

The primary barrier faced by those seeking disease freedom is the perceived financial burden associated with adopting biosecurity measures, as well as the challenges in comprehending and being uninformed of the OIE requirements. Peeler and Otte (2016) define disease freedom as the ultimate degree of protection that eradicates a significant portion of losses and expenses related to disease, such as heightened mortality, reduced growth, and medicine expenditures. The aquaculture business is always at a significant danger from a variety of pathogens, including bacteria, viruses, and parasites, which cause aquatic diseases. It is crucial to establish biosecurity programs that aim to prevent, contain, and eliminate pathogens from culture facility. If not properly controlled, certain illnesses may result in significant repercussions.

4.4.1 ECONOMIC LOSSES

Aquaculture enterprises might experience significant financial losses due to disease outbreaks. Profitability may be diminished by stock losses, decreased output, and the need for expensive treatments. Bacterial mortalities pose considerable obstacles to the development of aquaculture, leading to substantial economic losses (Singh et al., 2023; Madsen and Stauffer, 2024). The advancement of novel pharmaceuticals and immunizations serves as a supplementary means for controlling parasites, although their efficacy should be augmented by implementing other strategies such as biosecurity and cleanliness (Sitjà-Bobadilla and Oidtmann, 2017). Implemented strategies have the potential to mitigate the effects of bacterial and viral diseases by enhancing biosecurity measures. Yatabe et al. (2018) found that implementing biosecurity practices reduces the incidence of pathogens on farms. Additionally, Oidtmann et al. (2011) emphasized the significance of modeling disease introduction risk and conducting disease surveillance and monitoring in fish farms as crucial components of biosecurity practices.

4.4.2 ENVIRONMENTAL IMPACT

The evaluation of the consequences arising from aquaculture farm operations has not undergone a thorough examination that is sufficient to address the expansion of this business (Bohnes and Laurent, 2021). The potential ecological consequences of using antifouling paints, malachite green, and other chemical substances have not been well investigated. The environmental effects of a single farm may not be substantial when examined in isolation, but they may become considerable if additional farms, fishing areas, or activities are situated in the same vicinity. Due to the potential ecological impact, the release of viruses from aquaculture operations might disrupt wild aquatic species and their ecosystems. Effective environmental monitoring strategies facilitate the comprehension of waste management's effects, its utilization, and potential avenues for reduction. Thus, aquaculture operators must function as environmental stewards to guarantee a pristine environment for cultivating robust organisms (Carballeira Braña et al., 2021). The intake of nitrogen from salmonid farms leads to the creation of anoxic sediments and the emergence of toxic algal blooms in coastal waterways (Buschmann et al., 2006).

4.4.3 FOOD SECURITY

Aquaculture plays a crucial role in ensuring global food security and offers a viable solution to address the growing demand for aquatic food (Costello et al., 2020). The occurrence of disease outbreaks may cause disruptions in the consistent provision of fish and seafood, so affecting the accessibility of nourishing sustenance for millions of people. The provision of food, which is essential for the existence of humanity, is an ongoing and relevant topic. Therefore, due to aquaculture's distinct function, it is anticipated that the demand for aquatic goods would persistently rise as a result of exponential population expansion. Adequate comprehension of the biological and technological features of the cultivated species has been identified as a significant factor contributing to several achievements in aquaculture operations (Azra et al., 2021). Fish and other aquatic organisms obtained from marine and freshwater habitats are very nutritious animal-derived foods. Aquatic food has a high amount of easily absorbed minerals, including zinc, calcium, and long-chain n–3 polyunsaturated fatty acids. Consuming aquatic food is crucial for the development of the fetal brain and cognitive abilities, as well as for maintaining cardiovascular health in adults (Hibbeln et al., 2019; Garlock et al., 2022).

4.4.4 HUMAN HEALTH

As fish production continues to rise, a substantial quantity of trash is being created. It is crucial to implement effective treatment and management strategies to address the environmental and human health consequences of this waste. It is necessary to adopt regulations and laws (Ahmad et al., 2021). The proliferation of aquaculture exhibits spatial heterogeneity within a single country. The culture facilities are increasingly dependent on the use of formulated feeds, antibiotics, antifungals, and agrochemicals. As a result, the fish and shellfish harvested have higher levels of chemical and biological agents. These levels are higher compared to their wild counterparts (Sapkota et al., 2008). The use of nutritional, therapeutic, and preventive interventions in contemporary aquaculture methods may result in increased concentrations of antibiotic-resistant bacteria (in the water column, sediment, and bacterial strains linked with fish; Adams and Thompson, 2006), antibiotic residues, persistent organic pollutants, metals such as arsenic, methyl mercury, and lead, as well as parasites, and viruses, and parasites in aquacultured finfish and shellfish. Bacteria may disseminate mobile resistance genes and antibiotic-resistant genes more extensively by means of horizontal gene transfer, hence facilitating the spreading of genetic determinants. Upon activation, antibiotic resistance rapidly disseminates among aquatic microbial populations and thereafter may be transmitted to human pathogenic bacteria, rendering the usage of antibiotics ineffective for human health (Pepi and Focardi, 2021). The presence of heavy metals in water bodies and aquatic ecosystems has a significant impact on the food chain (Sonone et al., 2020). Since fish are consumed by people, their consumption indirectly impacts humans as well. Certain aquatic diseases possess zoonotic capacity, indicating their ability to transmit infections to people, hence presenting health hazards to those involved in or consuming aquaculture products.

4.5 MAJOR GOALS OF BIOSECURITY

4.5.1 ANIMAL MANAGEMENT

Ensuring the acquisition of robust stocks and enhancing their well-being and resistance via effective animal care, together with the enforcement of rigorous procedures for the examination and supervision of aquatic species, serves to avoid the introduction and dissemination of diseases that may cause significant harm to fish populations. Conducting routine health evaluations, implementing isolation protocols, and regulating the movement of animals inside aquaculture facilities are essential elements in ensuring the general welfare of aquatic species.

4.5.2 PATHOGEN MANAGEMENT

Ensuring the prevention, reduction, or elimination of infections is crucial in preserving biosecurity within aquaculture systems, since effective pathogen control is of utmost importance. Stringent sanitation protocols, frequent monitoring of water quality, and the use of resistant strains are necessary in order to avoid the growth of detrimental microorganisms. Disease transmission across farms has distinct variations and adheres to diverse patterns. Pathogens originating from upstream farms pose a threat to freshwater farmers. The transmission of this spread may be efficiently mitigated by implementing adequate physical distancing measures between farms. However, it will intensify as the number of farms in close proximity increases. Wild populations may transmit epidemic diseases to farmed fish populations. Disease may be transmitted across farms by either natural or human-caused carriers, so it can be well managed to ensure safety.

4.5.3 PEOPLE MANAGEMENT

Efficient management of personnel is crucial for preserving biosecurity in aquaculture operations. It is essential to provide training and education to managers and visitors on appropriate biosecurity-related policies, hygiene standards, and disease recognition. Enforcing rigorous protocols among personnel fosters a sense of accountability, guaranteeing a safe and protected setting. In 2018, the OIE created recommendations and best practices for public-private partnerships in the veterinary field (OIE, 2019). Additionally, they have launched an online training course to assist nations.

The facility's susceptibility to the entry and transmission of a particular pathogen, as well as the pathogen's ability to induce disease, determines the ease with which it may infiltrate the facility, propagate across different systems, and cause disease depend on:

- The farmed fish's species, immunological state, health, life stage, and sensitivity to strain.
- Key environmental elements include the quality and chemistry of water, as well as husbandry procedures adopted.
- Characteristics of the pathogen, including its biology, life cycle, possible reservoirs (carriers and other animals), capacity to survive on inanimate things (equipment), available legal treatment options, regulatory status (if it is an exotic or endemic disease, reportability, and compliance with local, state, and federal regulations).
- Workers' comprehension of biosecurity concepts and adherence to biosecurity standards.

4.6 COMPONENTS OF BIOSECURITY

The primary economic challenge in aquaculture arises during the grow-out phase, which is the longest period of the culture. This phase has a reduced likelihood of pathogen exclusion and accumulates a considerable portion of the investment over time, with feed expenses being the major cost. Utilizing contaminated fish diminishes the prospect of a successful harvest and fosters the proliferation of pathogens in the surroundings, hence perpetuating ailments in the region.

4.6.1 PHYSICAL BARRIERS

Physical barriers play a crucial role in maintaining biosecurity in aquaculture by acting as the primary defense against the entry and transmission of diseases. The barriers consist of robust fencing and enclosures that limit unauthorized entrance to aquaculture operations, hence limiting the introduction of possible disease vectors. In addition, the use of well-constructed infrastructure, such as input systems equipped with screens, aids in the filtration and prevention of diseases from infiltrating water sources utilized in aquaculture. Physical barriers are essential for separating aquatic creatures, limiting their interaction with wild populations, and decreasing the probability of cross-contamination. Regular examination and maintenance of these barriers is crucial to guarantee their efficacy in maintaining a bio-secure environment.

4.6.2 SANITATION AND DISINFECTION

Effective sanitation methods eliminate the organic residue that serves as a breeding ground for several disease-causing microorganisms. The efficacy of disinfectants is enhanced by removing mud and organic matter. Manual cleaning alone significantly decreases the quantity of contaminated organisms. A crucial sanitation measure involves ensuring the proper maintenance of aeration and filtration systems, as well as the appropriate removal and disposal of uneaten food, excess suspended particulates, dead or moribund fish (a significant source of pathogens), detritus, and additional organic matter, in accordance with relevant regulations. It is necessary to clean and subsequently disinfect all equipment, surfaces inside the facility, and vehicles/tires. Restricting vehicle access to certain regions is recommended, particularly if these are used for transportation to other aquaculture plants. It is crucial to comprehend that in some circumstances, achieving total disinfection may be unattainable; nonetheless, the use of disinfectants will significantly decrease the quantity of pathogens. Several disinfection techniques are available for use in aquaculture. The effectiveness of different approaches in a certain context will be determined by logistics, residual toxicity, and cost. The majority of disinfection procedures may be categorized as either chemical or physical. Aquaculturists must acquaint themselves with the federal, state, and municipal rules pertaining to the use of a certain technique or substance, as well as any potential risks to human health and the environment. Physical techniques for disinfecting equipment comprise heat, direct sunlight, and drying. Exposing microorganisms to temperatures ranging from 80 to 100°C for duration of 10 min effectively eliminates all active germs. However, certain life stages, such as spores of bacteria, may need much longer exposure times. The efficacy of sunlight varies based on elements such as intensity, temperature, and other variables. However, there is little information available on the optimal duration of exposure. The use of drying equipment may effectively decrease the population of pathogens, while it is possible for spores, eggs, or cysts to remain viable after the treatment.

4.6.3 QUARANTINE PROCEDURES

Quarantine methods are crucial for maintaining biosecurity in aquaculture by avoiding the introduction of microbes to existing populations, since eradicating them once they have entered the facility would be challenging or impossible. Aquatic organisms that have just been obtained undergo a certain period of isolation, during which they are closely observed and tested in order to detect any possible disease-causing agents. Amidst the quarantine period, rigorous biosecurity measures, such as isolating from existing populations, effectively lessen the chances of disease spread. The duration of quarantine should consider data on incubation periods and development timeframes for the infectious agents known to pose a threat. The period of quarantine might vary from 15 days to three months (Jia et al., 2017) in order to prevent the emergence of a potentially severe "new" issue. Once a condition has been accurately diagnosed, therapy should be administered using effective medications for the right duration. Prophylactic medicines may prevent the development of clinical symptoms, but the improper administration should not be practiced (Adams and Thompson, 2006). The implementation of quarantine measures might incur significant expenses, so it is advisable to take them into account at the outset of the facility design process. Regular health evaluations and diagnostic tests are performed to verify the health condition of confined organisms prior to their introduction to the primary aquaculture unit. Efficient quarantine protocols function as a defensive strategy, ensuring the general well-being and security of aquaculture operations.

4.6.4 PATHOGEN DETECTION AND MONITORING

Diagnostic testing is employed to identify the occurrence and monitor the advancement of infectious organisms in fish populations. These tests are commonly utilized to achieve four primary biosecurity-related objectives. These objectives include demonstrating the absence of infection in aquaculture in order to obtain or maintain infection-free certification, screening fish prior to their introduction to a new facility, identifying fish that are affected as soon as feasible during a time of quarantine, and confirming suspicious or clinical cases (Ninawe et al., 2017). Progress in molecular biology has facilitated enhanced comprehension of pathogen biology, pathogenicity, and behavior, leading to improved and expedited diagnostic processes and testing. Sequencing-based molecular diagnostic methods are often used for the identification of nucleic acids (DNA, RNA) in molecular detection. Utilizing eDNA and eRNA may ascertain the presence of pathogens in the aquatic environment (Chouhan et al., 2023), hence reducing the need for physically demanding sample procedures.

4.6.5 USE OF SPR, SPF, AND SPT

The establishment of domesticated specific pathogen-free (SPF) populations of *Litopemeus vennamei* has facilitated the implementation of biosecure shrimp farming practices. It has been unequivocally demonstrated that the development, maintenance, and utilization of domesticated SPF shrimp populations significantly diminishes the likelihood of disease outbreaks in shrimp aquaculture. Moreover, this approach enables the intensification of production and the subsequent enhancement of profitability. However, SPF stocks may still contain certain infections; thus, it is crucial to have highly biosecure facilities that adhere to strict biosecurity management protocols. Specific pathogen tolerant (SPT)/specific pathogen resistant (SPR) animals are chosen to establish SPF stock for utilization in a low biosecure setting. SPF stocks are manufactured using a rigorous monitoring system that includes precise diagnostic procedures and provide several benefits compared to non-SPF stocks (Eswaran, 2022). They are raised in places where particular harmful pathogens are deliberately kept out. It is still difficult to make genetically modified seed affordable and acceptable for smallholder farmers since SPR, SPT, and other genetically enhanced seeds are somewhat more expensive than locally accessible seed of the target species. Advancements in genetics and molecular biology provide significant possibilities for disease management, either via selective breeding for disease resistance or through the implementation of efficient and precise culture site testing.

4.7 HEALTH CERTIFICATION

Aquaculture health certification entails a comprehensive evaluation of several criteria, such as the quality of water, the control of diseases, and the general welfare of aquatic species. Regulatory organizations often provide these certificates to demonstrate that aquaculture operations comply with prescribed health requirements. Through the establishment and upkeep of health certification in aquaculture, producers may bolster the legitimacy of their goods in the market, providing customers with confidence in the safety and quality of the seafood they eat. Furthermore, the implementation of health certification will play a crucial role in preventing and controlling diseases, hence reducing the potential harm caused by outbreaks to both aquatic environments and the aquaculture business. The certification process promotes sustainable practices and cultivates environmental responsibility within the sector. Additionally, it streamlines international commerce by guaranteeing that aquaculture products adhere to the health and safety standards set by the nations that import them. Health certification in aquaculture encompasses a comprehensive strategy that fosters conscientious and secure methodologies, safeguards the well-being of aquatic organisms, and enhances the worldwide seafood distribution network. An effective and economical method to prevent the introduction of diseases to a farm is to guarantee that only animals free from infectious pathogens are permitted into the premises. In this scenario, "health certificate" is used to verify that the fish population has undergone testing and is guaranteed to be devoid of any diseases or contagious infections. The issuance of certificates is a straightforward procedure that fully complies with the fundamental criteria set by OIE regulations. No organization has officially implemented the biosafety certification system that prevents the spread of diseases. However, there is a certification available for aquaculture farming operations that ensures freedom from diseases (Scarfe and Palic, 2020). To effectively manage diseases and invasive pests, it is essential to establish comprehensive strategies that include all nations within the area.

4.8 PRACTICES INVOLVED IN MAINTAINING POND BIOSECURITY

Essential biosecurity procedures include hazard identification, risk assessment, risk management, effective quarantine of stocks, isolation of affected stocks, maintenance of personal hygiene, and control of humans, animals, and vectors.

4.8.1 FARM LAYOUT AND DESIGN

An intelligently planned farm structure and design are crucial in enhancing biosecurity measures in aquaculture operations. Strategically positioning ponds, raceways, and other infrastructure may effectively reduce the likelihood of disease entry and transmission. Enforcing one-way movement of water helps prevent the spread of contamination in different areas. Optimal distance between ponds and well-defined isolation zones is crucial for implementing successful quarantine measures, therefore mitigating the transmission of disease. In addition, the implementation of regulated access points equipped with biosecurity measures, such as footbaths and disinfection stations, decreases the probability of unauthorized individuals bringing infections. Effective waste management methods also reduce the potential for waterborne contamination. Well-designed facilities also enable the enforcement of stringent hygiene rules for equipment, hence minimizing the risk of cross-contamination. Essentially, a well-structured arrangement and design of a farm play a crucial role in creating a biosecure habitat that is necessary for the sustained and healthy growth of aquatic species. When choosing a location for a new shrimp growing facility, it is important to find a region where certain illnesses are not often found (Lightner, 2005).

4.8.2 EMPLOYEE TRAINING

A thorough personnel training is crucial for strengthening biosecurity controls in aquaculture operations. Staff members who possess extensive knowledge are more capable of identifying and dealing with possible biosecurity concerns such as limited access zones and quarantine measures, hence reducing the likelihood of disease introduction. Training programs should prioritize the significance of rigorous hygiene practices, appropriate disinfection processes, and the use of protective gear to effectively avoid the inadvertent transmission of diseases. Moreover, training promotes the development of a culture that emphasizes being conscious and responsible for maintaining a biosecure environment. By adopting this proactive strategy, the chances of infections being transmitted by anthropogenic intervention are greatly reduced, hence enhancing the general health and future viability of aquaculture operations.

4.8.3 STOCK SOURCING

Shrimp producers are increasingly choosing to stock their ponds with post larvae that come from healthy, captive-bred broodstock. They do this to reduce the risk of crop loss caused by disease (Lightner et al., 2009). Controlling the disease status of captive broodstock may be achieved to a considerable degree by using shrimp that are free from certain pathogens (SPF) (Lightner et al., 2009). SPF stocks specifically refer to animals that have been kept at the breeding center and have previously completed thorough quarantine and screening procedures. After departing from the breeding center, stocks are classified as high-health, indicating that they are devoid of certain infections (Pruder, 2004). SPF shrimp are devoid of one or more certain diseases. Farms that had a higher biosecurity score showed a correlation with a reduced number of different pathogens present. On the other hand, farms that had an in-degree greater than one, meaning they had more than one fish supplier, showed a correlation with an increased number of different pathogens present (Yatabe et al., 2018).

4.8.4 FEED AND WATER MANAGEMENT

Efficient control of feed and water is crucial for improving biosecurity measures in aquaculture operations. Thorough surveillance and regulation of feed sources may reduce the probability of introducing harmful substances that might jeopardize the well-being of aquatic organisms. Adopting stringent storage and handling protocols for feed minimizes the risk of contamination and transmission of disease-causing microorganisms. In addition, implementing a rigorous water quality management system guarantees a hygienic and pathogen-free habitat for aquatic creatures. Consistent surveillance of water sources, along with effective filtration and disinfection protocols, aids in the prevention of the introduction and further spread. By giving priority to these factors, aquaculture facilities may take proactive measures to prevent the entry of hazardous agents, ensuring a biosecure environment that is crucial for the optimum development and well-being of produced species. The use of unprocessed or untreated water is accountable for the transmission of diseases when ponds are populated with high-health fish and shrimp offspring. Proper handling and storage may minimize the presence of foodborne pathogens. To reduce the chance of pests and diseases entering via incoming water, it is important to maintain appropriate treatment measures.

4.8.5 SURVEILLANCE FOR DISEASES OF FISH IN AQUACULTURE

Without excellent health data, it is impossible to design any aquatic health strategy or policy for aquatic animal health. This data may be used for the purposes of disease management, quarantine, and health certification via the implementation of aquatic animal monitoring (Lee et al., 2008). A key component of any biosecurity plan is surveillance to prevent the introduction of disease. By identifying potential pathways for disease introduction into aquatic environments and spotting the emergence of new diseases, control measures can be put in place before the pathogen spreads widely (Oidtmann et al., 2011). Regular monitoring is crucial for mitigating the risk of pathogen transmission (Oidtmann et al., 2011). Disease monitoring should be an inherent and vital aspect of government services for the health of aquatic animals (Hadfield and Clayton, 2011). Pathogen richness determinants might be considered as factors to include in risk-based surveillance programs (Oidtmann et al., 2013).

4.8.6 *EMERGENCY RESPONSE AND DISEASE CONTROL*

An effective emergency response and disease management plan are of utmost importance in rigorous biosecurity policies used on aquaculture operations. Prompt identification and segregation of afflicted individuals, together with rapid reporting, are essential for limiting possible epidemics. Enacting predetermined emergency protocols enables a swift and systematic reaction, hence reducing the transmission of diseases. Consistent training exercises guarantee that farm staff are adequately equipped to implement biosecurity protocols under crucial circumstances. By using sophisticated diagnostic tools, the farm's capacity to swiftly detect diseases increased, thereby ensuring the overall health and long-term viability of the aquaculture enterprise. Significant advancements have been made in comprehending and managing diseases like *Aeromonas salmonicida* and *Yersinia ruckeri* in specific cultured species through the development of treatment strategies through vaccination. Vaccination and treatment are crucial measures that can be employed to effectively address and control disease outbreaks, while also improves the overall health of cultured fish.

4.9 PROFESSIONALIZING BIOSECURITY

The systematic integration of best practices and established standards for the enhancement of overall health is known as professionalizing biosecurity. The process starts with extensive training programs designed for farm staff, guaranteeing a solid grasp of biosecurity concepts and practices. A secure environment is established by the adoption of rigorous access restrictions, such as restricted entrance points and comprehensive disinfection measures. Consistent health monitoring and disease surveillance facilitate the timely identification and swift action. To enhance the professionalism of these efforts, it is crucial to hire competent biosecurity officers and allocate resources toward new diagnostic technology. Engaging with regulatory agencies and industry experts nurtures a culture of ongoing enhancement and compliance with global biosecurity standards. By placing professionalism as a top priority in implementing biosecurity measures, culture facilities may effectively reduce the dangers of disease while also fostering ethical and sustainable practices in this business.

4.10 FUTURE ACTIVITIES AND THE WAY FORWARD

The future will see an increase in the risk of disease introductions due to the inevitable growth in international commerce of animal products and humans, as well as the expansion of aquaculture into new areas. Extension and researcher personnel bear the duty of forming alliances with farmers to conduct essential research and utilize the findings to develop an economically efficient strategy. Risk analysis has the dual purpose of formulating biosecurity policies and pinpointing significant deficiencies in current knowledge, so shaping the trajectory of future research. The objective is to create and implement national, regional, and inter-regional policies and regulatory frameworks that are suitable and efficient in managing the introduction and transportation of live aquatic animals and products. Enhance the capabilities of institutions and farmers through education and extension to promote capacity building. Creating and executing efficient national disease reporting systems, databases, and other methods for gathering and examining information on aquatic animal diseases. By validating precise and sensitive diagnostic techniques, safe therapeutic agents, efficient disease control strategies, and investigations into new diseases and infections. A comprehensive strategy for managing the health of aquatic animals with a focus on preventive interventions and maintenance of favorable environments should be advocated. Exploring alternative approaches to health management, such as employing disease-resistant, domesticated varieties of aquatic animals, in order to mitigate the effects of diseases. Additional research endeavors are necessary to create health management tools and techniques for economically viable species, such as carps, tilapias, and catfishes, which are predominantly produced by small-scale farmers. However, the process of implementing new information and technology on a worldwide scale has been hindered by insufficient distribution and acceptance of these advancements at the local level. The increased adoption and utilization of SPF, SPR, and genetically modified seeds, driven by the demonstration of substantial economic advantages, will enhance their usage among smallholder farmers in the future years.

4.11 CONCLUSION

Farm-level biosecurity measures encompass a range of activities, including implementing rigorous quarantine protocols, utilizing disinfectants and pesticides, disinfecting fish seed, treating water, using uncontaminated feed, and correctly disposing of dead animals. Occasionally, farmers perceive these measures as an unnecessary financial tension, failing to recognize their potential beneficial effects. The majority of small and marginal farmers in developing nations do not adhere to any biosecurity precautions on their farms. Poor animal care practices can activate secondary (opportunistic) pathogens, leading them to become more harmful and produce a higher number of diseases. Therefore, it is crucial to implement comprehensive measures to avoid the introduction of disease-causing agents. To create a standardized and cost-effective product on demand, it is essential to integrate biosecurity, health, nutrition, genetics, and environmental quality. From an aquaculture perspective, the primary goal is to mitigate the economic consequences of disease. This entails establishing a conceptual framework to develop biosecurity programs that aim to limit the prospect of disease introduction, and in essence, biosecurity serves as a tool for promoting sustainability beyond horizons.

KEYWORDS

- **aquaculture**
- **biosecurity**
- **disease**
- **quarantine**
- **monitoring**
- **health certification**

REFERENCES

Adams, A., & Thompson, K. D. (2006). Biotechnology offers revolution to fish health management. *Trends in Biotechnology*, *24*(5), 201–205.

Ahmad, A., Abdullah, S. R. S., Hasan, H. A., Othman, A. R., & Ismail, N. I. (2021). Aquaculture industry: Supply and demand, best practices, effluent and its current issues and treatment technology. *Journal of Environmental Management*, *287*, 112271.

Azra, M. N., Okomoda, V. T., Tabatabaei, M., Hassan, M., & Ikhwanuddin, M. (2021). The contributions of shellfish aquaculture to global food security: Assessing its characteristics from a future food perspective. *Frontiers in Marine Science*, *8*, 654897.

Birhan, G., Alebie, A., Admassu, B., Shite, A., Mohamed, S., & Dagnaw, B. (2015). A review on emerging and reemerging viral zoonotic diseases, *International Journal of Basic and Applied Virology*, *4*(2), 53–59.

Bohnes, F. A., & Laurent, A. (2021). Environmental impacts of existing and future aquaculture production: Comparison of technologies and feed options in Singapore. *Aquaculture*, *532*, 736001.

Buschmann, A. H., Riquelme, V. A., Hernández-González, M. C., Varela, D., Jiménez, J. E., Henríquez, L. A., Vergara, P. A., Guíñez, R., & Filún, L. (2006). A review of the impacts of salmonid farming on marine coastal ecosystems in the southeast Pacific. *ICES Journal of Marine Science*, *63*(7), 1338–1345.

Carballeira Braña, C. B., Cerbule, K., Senff, P., & Stolz, I. K. (2021). Towards environmental sustainability in marine finfish aquaculture. *Frontiers in Marine Science*, *8*, 343.

Chouhan, N., Dekari, D., Choudhary, B., Singh, A., & Choudhury, T. G. (2023). Environmental DNA (eDNA) technology: Fisheries and aquaculture perspectives. *Indian Journal of Animal Health*, *62*, 2.

Costello, C., Cao, L., Gelcich, S., Cisneros-Mata, M. Á., Free, C. M., Froehlich, H. E., Golden, C. D., Ishimura, G., Maier, J., Macadam-Somer, I., & Mangin, T. (2020). The future of food from the sea. *Nature*, *588*(7836), 95–100.

Eswaran, S. (2022). Specific pathogen free (SPF) shrimps in aquaculture. In: *Advances in Fisheries Biotechnology*. Springer Nature Singapore: Singapore, pp. 465–470.

FAO, IFAD, UNICEF, WFP and WHO. (2018). *The State of Food Security and Nutrition in the World 2018: Building Climate Resilience for Food Security and Nutrition.* Food and Agriculture Organization of the United Nations (FAO), Rome, Italy, the International Fund for Agricultural Development (IFAD), the United Nations Children's Fund (UNICEF), the World Food Programme (WFP) or the World Health Organization (WHO): Rome, Italy, p. 181.

Garlock, T., Asche, F., Anderson, J., Ceballos-Concha, A., Love, D. C., Osmundsen, T. C., & Pincinato, R. B. M. (2022). Aquaculture: The missing contributor in the food security agenda. *Global Food Security*, *32*, 100620.

Hadfield, C. A., & Clayton, L. A. (2011). Fish quarantine: Current practices in public zoos and aquaria. *Journal of Zoo and Wildlife Medicine*, *42*(4), 641–650.

Hibbeln, J. R., Spiller, P., Brenna, J. T., Golding, J., Holub, B. J., Harris, W. S., Kris-Etherton, P., Lands, B., Connor, S. L., Myers, G., & Strain, J. J. (2019). Relationships between seafood consumption during pregnancy and childhood and neurocognitive development: Two systematic reviews. *Prostaglandins, Leukotrienes and Essential Fatty Acids*, *151*, 14–36.

Jia, B., St-Hilaire, S., Singh, K., & Gardner, I. A. (2017). Biosecurity knowledge, attitudes and practices of farmers culturing yellow catfish (Pelteobagrus fulvidraco) in Guangdong and Zhejiang provinces, China. *Aquaculture*, *471*, 146–156.

Laltlanmawia, C., Ghosh, L., Saha, R. K., Parhi, J., Pal, P., Dhar, B., & Saha, H. (2023). Isolation, identification and pathogenicity study of emerging multi-drug resistant fish pathogen Acinetobacter pittii from diseased rohu (Labeo rohita) in India. *Aquaculture Reports*, *31*, 101629.

Lee, C. S., Scarfe, A. D., & O'Bryen, P. J. (2008). *Aquaculture Biosecurity: Prevention, Control, and Eradication of Aquatic Animal Disease*. John Wiley & Sons.

Lightner, D. V. (2005). Biosecurity in shrimp farming: Pathogen exclusion through use of SPF stock and routine surveillance. *Journal of the World Aquaculture Society*, *36*(3), 229–248.

Lightner, D. V., Redman, R. M., Arce, S., & Moss, S. M. (2009). Specific pathogen-free shrimp stocks in shrimp farming facilities as a novel method for disease control in crustaceans. In: *Shellfish Safety and Quality*. Woodhead Publishing, pp. 384–424.

Madsen, H., & Stauffer Jr, J. R. (2024). Aquaculture of animal species: Their eukaryotic parasites and the control of parasitic infections. *Biology*, *13*(1), 41.

Ninawe, A. S., Hameed, A. S., & Selvin, J. (2017). Advancements in diagnosis and control measures of viral pathogens in aquaculture: An Indian perspective. *Aquaculture International*, *25*, 251–264.

Oidtmann, B., Peeler, E., Lyngstad, T., Brun, E., Jensen, B. B., & Stärk, K. D. (2013). Risk-based methods for fish and terrestrial animal disease surveillance. *Preventive Veterinary Medicine*, *112*(1–2), 13–26.

Oidtmann, B. C., Thrush, M. A., Denham, K. L., & Peeler, E. J. (2011). International and national biosecurity strategies in aquatic animal health. *Aquaculture*, *320*(1–2), 22–33.

OIE. (2019). *Recommendations from the 4th OIE Global Conference on Aquatic Animal Health: Collaboration, Sustainability-Our Future*. April 2–4, 2019, World Organization for Animal Health (OIE): Santiago, Chile. Paris, France, p. 5.

Paul, A., Mohanty, S., Parida, S., Biswal, S., Sandha, B. K., & Sahoo, P. K. (2023). Tilapia parvovirus disease: An emerging threat for the tilapia aquaculture industry.

Peeler, E. D. M. U. N. D. (2005). The role of risk analysis and epidemiology in the development of biosecurity for aquaculture. In *Diseases in Asian Aquaculture V. Fish Health Section, Asian Fisheries Society, Manila, Philippines, Queensland, Australia*, pp. 35–46.

Peeler, E. J., & Otte, M. J. (2016). Epidemiology and economics support decisions about freedom from aquatic animal disease. *Transboundary and Emerging Diseases*, *63*(3), 266–277.

Pepi, M., & Focardi, S. (2021). Antibiotic-resistant bacteria in aquaculture and climate change: A challenge for health in the Mediterranean area. *International Journal of Environmental Research and Public Health*, *18*(11), 5723.

Preena, P. G., Dharmaratnam, A., Raj, N. S., Raja, S. A., Nair, R. R., & Swaminathan, T. R. (2021). Antibiotic-resistant Enterobacteriaceae from diseased freshwater goldfish. *Archives of Microbiology*, *203*(1), 219–231.

Pruder, G. D. (2004). Biosecurity: Application in aquaculture. *Aquacultural Engineering*, *32*(1), 3–10.

Saegerman, C., Dal Pozzo, F., & Humblet, M. F. (2012). Reducing hazards for humans from animals: Emerging and re-emerging zoonoses. *Italian Journal of Public Health*, *9*(2), 13–24.

Sapkota, A., Sapkota, A. R., Kucharski, M., Burke, J., McKenzie, S., Walker, P., & Lawrence, R. (2008). Aquaculture practices and potential human health risks: Current knowledge and future priorities. *Environment International*, *34*(8), 1215–1226.

Scarfe, A. D., & Palić, D. (2020). Aquaculture biosecurity: Practical approach to prevent, control, and eradicate diseases. In: *Aquaculture Health Management*. Academic Press, pp. 75–116.

Singh, A., PavanKalyan, M., Choudhury, T. G., Kamilya, D., Khan, M. I. R., & Chouhan, N. (2023). Supplementation of autochthonous potential probiotic Bacillus subtilis COFCAU_BSP3 to Labeo rohita feed: Effect on immune-biochemical responses and resistance against *Aeromonas hydrophila*. *Aquaculture International*, *32*, 1–16.

Sitjà-Bobadilla, A., & Oidtmann, B. (2017). Integrated pathogen management strategies in fish farming. In: *Fish Diseases*. Academic Press, pp. 119–144.

Sonone, S. S., Jadhav, S., Sankhla, M. S., & Kumar, R. (2020). Water contamination by heavy metals and their toxic effect on aquaculture and human health through food Chain. *Letters in Applied NanoBioScience*, *10*(2), 2148–2166.

Stentiford, G. D., Bateman, I. J., Hinchliffe, S. J., Bass, D.1., Hartnell, R., Santos, E. M., Devlin, M. J., Feist, S. W., Taylor, N. G. H., Verner-Jeffreys, D. W., & Van Aerle, R. (2020). Sustainable aquaculture through the One Health lens. *Nature Food*, *1*(8), 468–474.

Subasinghe, R., & Shinn, A. P. (2023). Biosecurity: Current and future strategies. In: *Climate Change on Diseases and Disorders of Finfish in Cage Culture*. CABI: GB, pp. 430–461.

Subasinghe, R., Alday-Sanz, V., Bondad-Reantaso, M. G., Jie, H., Shinn, A. P., & Sorgeloos, P. (2023). Biosecurity: Reducing the burden of disease. *Journal of the World Aquaculture Society*, *54*(2), 397–426.

Subasinghe, R., Soto, D., & Jia, J. (2009). Global aquaculture and its role in sustainable development. *Reviews in Aquaculture*, *1*(1), 2–9.

Subasinghe, R. P., & Bondad–Reantaso, M. G. (2005). Biosecurity in aquaculture: International agreements and instruments, their compliance, prospects, and challenges for developing countries. In: Scarfe, D., Lee, C-S., O'Bryen, P. J., *Aquaculture Biosecurity: Prevention, Control, and Eradication of Aquatic Animal Disease*, Wiley, pp. 9–16.

van Senten, J., Engle, C. R., Hartman, K., Johnson, K. K., & Gustafson, L. L. (2018). Is there an economic incentive for farmer participation in a uniform health standard for aquaculture farms? An empirical case study. *Preventive Veterinary Medicine*, *156*, 58–67.

Yatabe, T., More, S. J., Geoghegan, F., McManus, C., Hill, A. E., & Martinez-Lopez, B. (2018). Can biosecurity and local network properties predict pathogen species richness in the salmonid industry. *PLoS One*, *13*(1), e0191680.

CHAPTER 5

Nutrigenomics: Molecular Nutrition Strategies for Aquaculture Enhancement

VIVEK CHAUHAN[1], ARYA SINGH[2], and NAYAN CHOUHAN[3*]

[1]*Faculty of Fisheries Science, Kerala University of Fisheries and Ocean Studies, Kochi, Kerala, India*

[2]*College of Fisheries, Acharya Narendra Deva University of Agriculture and Technology, Ayodhya, India*

[3]*College of Fisheries, Central Agricultural University, Lembucherra, Tripura, India*

*Corresponding author

ABSTRACT

Nutrigenomics aims to investigate the comprehensive effects of nutrition on the genome and the groundbreaking overlap of genomics and nutrition to understand the intricate interactions that affect the performance and health of aquatic species. Nutrigenomics investigation offers a thorough comprehension of how dietary components impact gene expression, providing exceptional insights into customized and precise nutrition recommendations. This chapter presents the fundamental concepts, methodology, and possible uses of nutrigenomics, highlighting its importance in improving feed formulations, boosting growth performance, strengthening disease resistance, and promoting general well-being in cultured aquatic species. It is well recognized that there are variations in dietary needs across various animals. Therefore, the investigation of nutrigenomics in various fish and shellfish species has great significance. This chapter also elucidates the genetic basis of the complex interplay between nutrition and genetics,

Advances in Freshwater Aquaculture: Technology for Sustainable Development. Tarang Kumar Shah, Madhu Sharma, Mohd. Ashraf Rather, & Avdhesh Kumar (Eds.)

DOI: 10.1201/9781779640857-5

paving the way for a revolutionary period in aquaculture characterized by accuracy, sustainability, and the exploration of new prospects for the sector.

5.1 INTRODUCTION

In the ever-changing field of aquaculture, where maximizing productivity and ensuring the welfare of aquatic species are of utmost importance, it is crucial to use advanced scientific methods. Out of these techniques, nutrigenomics is a remarkable field that has the potential to uncover the complex relationship between nutrition, genetics, and physiological responses of aquatic species. The overall physiological functioning and health of an organism are determined by the interplay of nutrition, environment, and genome. The field of study that seeks to comprehend these relationships is referred to as "Nutrigenomics." Aquatic nutrigenomics is a field of study that investigates the relationship between dietary components and gene expression in aquatic creatures, with the aim of understanding how this interaction affects their performance and health. In the pursuit of sustainability, efficiency, and environmental responsibility, the aquaculture sector is investigating nutrigenomics to gain valuable knowledge in developing customized and precise feeding strategies for various species.

Gene expression research in aquaculture has advanced from traditional single-gene sequencing to contemporary sequencing methods, including nutritional genomics, metagenomics, epigenetics, and several other omics techniques. Metagenomics research has facilitated the prompt identification of diseases in aquatic animals by accelerating the detection of newly developing diseases (Arshad et al., 2017). Nutrigenomics, also referred to as "nutritional genomics," has significantly enhanced our understanding of molecular markers for nutrition-related diseases. Additionally, it has expanded our ability to promote optimal nutrition requirements that can enhance growth performance, lipid metabolism, reproductive development, and maintain a stable immune system in various aquatic organisms.

Nutrigenomics is a scientific method that combines nutritional sciences with genomics that utilizes molecular techniques to investigate, evaluate, and comprehend the many reactions elicited by a certain diet in individuals or groups of people (Craig et al., 2017). It also examines the broad-ranging effects of nutrition on the whole genome. From a nutrigenomics standpoint, nutrients serve as dietary signals that are recognized by cellular sensor systems, which in turn alter gene and protein expression, leading to the generation of metabolites. In the field of fish nutrition, particularly for species raised in aquaculture, the composition of formulated feeds differs greatly from their natural diet (DeBusk, 2009). There have been recent shifts in the sources of nutrients, particularly protein and oil, which are now mostly supplied from terrestrial materials. In addition, health management now includes the use of functional diets, which are considered to enhance fish health, decrease disease outbreaks, and promote recovery after illness. Fish are much more exposed to pathogens compared to land-dwelling vertebrates. In saltwater, there are usually 1 million bacteria and 10 million viruses per milliliter, including both pathogenic and nonpathogenic varieties (Drayna, 2005). This technology involves using advanced technologies like transcriptomics, proteomics, and metabolomics to study how diet affects health (Fuhrman, 1999).

The growing aquaculture sector necessitates a corresponding rise in the manufacturing of aquafeeds. According to Gardy and Loman (2018), fish meal and fish oil are the costliest protein sources used in animal and aquaculture diets. The price of fish meal has been steadily rising, making it the most costly protein source and leading to a dramatic rise in the price of commercial feed. Efforts have been made to replace fish meal with other protein sources. Substituting plant items for marine resources in fish diets adversely affects the quality, health, and development of farmed fish. The importance of nutrigenomics in improving feed formulations, enhancing growth performance, boosting disease resistance, and promoting the general well-being of aquaculture species will become more apparent over time. By deciphering the genetic codes that regulate the delicate interplay between nutrition and genetics, we are laying the foundation for a new age in aquaculture—a future where accuracy and sustainability are not just goals, but attainable certainties. This chapter provides an introduction to the study of nutrigenomics in the field of aquaculture. It explains the fundamental ideas, methods, and possible uses of this field of research. This will provide them with a thorough comprehension of how genetic information and nutritional variables combine to influence the results of aquaculture operations.

5.2 IMPORTANCE OF NUTRIGENOMICS IN AQUACULTURE

Fisheries and aquaculture are fast-growing and remain a vibrant and important production sector for high-protein food, having an important role to play in achieving food and nutritional security. However, like other food production sectors, it also faces numerous challenges. The rapid expansion of global aquaculture production simultaneously increases the demand for aquaculture feeds.

Feed provides the nutrients required for building and sustaining life. Understanding the nutritional requirements and need-based dietary manipulation is perhaps the most effective and sustainable way of dealing with these challenges. In this review, we discuss the concept of nutrigenomics as a subset of multiomics technology and highlight some studies encompassing nutrigenomics studies, with a specific focus on its application in fish nutrition. Partial replacement of fish meal with other plant proteins resulted in excellent growth rates but total replacement of fish meal by plant proteins seems to be linked to lower growth performance in some fishes. Further investigations in carnivorous fish species prove that they do not use dietary carbohydrates efficiently. Therefore, it is essential to provide a greater understanding of the use of nutrients and regulation of metabolism in fish nutrition. A further approach in fish nutrition is a thorough analysis of genes that encode these enzymes using recent developments in genomic technology.

Various studies have discovered a strong correlation between the positive impacts of providing appropriate nutrition and the efficient achievement of production, development, sexual maturity, molting, antioxidant capacity, coloration in ornamental fishes, and reproduction in desirable aquatic species. These benefits may be successfully boosted by including certain nutrients. Furthermore, the modulation of gene expression has the potential to enhance immunity as well. Nutrigenomics has had significant growth in this field, focusing on many areas of the impact of nutrients on development.

5.3 BASICS OF NUTRIGENOMICS

5.3.1 OVERVIEW OF NUTRIGENOMICS

The application of nutrigenomics in aquaculture holds the promise of promoting sustainability and efficiency in the industry, offering a pathway to address challenges related to environmental impact, resource utilization, and the overall well-being of aquatic species. Nowadays, the "Post-Genomic Era" is being experienced. This era is characterized by the integration of three fields: biological, social, and environmental, where scientific discoveries on nutritional pathophysiology and metabolism are included. The interaction between genetic and dietary influences can result in a higher risk of disease in certain individuals and populations. Researches indicate that diet–gene interactions play a significant role in person variability and have clarified some of these genetic differences (Ghoshal et al., 2003). Variability between individuals in response to dietary intervention is a well-known phenomenon in nutrition research and practice (Holliday et al., 1993). In the 20th century, nutritional science primarily concentrated on identifying vitamins (Lende et al., 2014) and minerals (Martin and Król, 2017), to address nutritional deficiencies and prevent the onset of diseases. Currently, nutrition research is exploring how nutrition can optimize and sustain cellular, tissue, organ, and overall body homeostasis. To comprehend this, one must have knowledge of how nutrients function on a molecular scale. This, in turn, entails a complex web of interactions between nutrients and genes, proteins, and metabolic processes. The arrangement of gene expression, protein expression, and metabolite creation in response to certain nutrients or nutritional patterns might be regarded as "dietary signatures." Nutrigenomics aims to analyze the unique nutritional patterns inside individual cells, tissues, and organisms and to comprehend the impact of nutrition on maintaining internal stability. In addition, nutrigenomics seeks to uncover the genes that impact the likelihood of diet-related disorders on a comprehensive genomic level and to comprehend the processes that underlie these genetic predispositions.

5.3.2 PRINCIPLES OF NUTRIGENOMICS

Nutrigenomics elucidates the genetic mechanisms via which typical food constituents modify the equilibrium between well-being and illness by modifying the expression and/or configuration of an individual's genetic constitution. In the field of aquaculture, the fundamental concepts of nutrigenomics provide a revolutionary method for enhancing the well-being and efficiency of aquatic creatures. This encompasses a wide range of research approaches, spanning from fundamental studies in cellular and molecular biology to clinical trials, epidemiology, and population health. A gene is the fundamental and tangible entity of heredity that is transmitted from one generation to the next. Genes are discrete sections of DNA that include the instructions necessary for synthesizing a particular protein. Various genetic variants exist, including single-nucleotide polymorphisms (SNPs), which are changes in a single nucleotide. Nutritional genomics is the field of study that use molecular techniques to investigate, evaluate, and comprehend the many physiological reactions elicited by a certain dietary regimen in individuals or groups of people (Menon et al., 2010). Nutrigenomics is the study of how the genetic material of aquatic animals interacts with food components. The underlying premise is to acknowledge that individual genetic differences have an impact on how organisms react to various nutrients. This observation enables the adaptation of feed compositions, considering the genetic variability among populations. The concepts also apply to the regulation of gene expression, where certain nutrients influence the activation or inhibition of genes linked to important characteristics such as growth, disease resistance, and general health. Customizing diets according to an individual's genetic composition guarantees a more accurate and efficient approach to nutrition. It highlights the shift from a standardized, one-size-fits-all strategy to a more personalized and species-specific manner. Furthermore, the ideas of nutrigenomics in aquaculture emphasize the ever-changing nature of these interactions. They emphasize the significance of continuous study to untangle the intricacies of gene–nutrient interactions, consistently improving our comprehension of how to improve feed compositions for various species and environmental circumstances. Aquaculturists may optimize growth efficiency, disease resistance, and general sustainability in the sector by following these principles and using nutrigenomics.

5.3.3 GENETIC AND EPIGENETIC FACTORS IN AQUACULTURE

Nutrigenomics is a strong tool in the ever-changing field of aquaculture, where the complex interaction between genetic and epigenetic variables determines how aquatic species respond to their nutrition. The genetic elements encoded in an organism's DNA determine fundamental characteristics, including the speed at which it grows, its ability to withstand diseases, and how efficiently it utilizes nutrients. Nutrigenomics explores the genetic markers to develop customized nutrition strategies that enhance the development of advantageous characteristics in aquaculture species. Concurrently, the impact of epigenetic variables adds an additional level of intricacy to the interaction between nutrition and heredity. Epigenetic alterations, such as DNA methylation and histone modifications, have the ability to modify gene expression without modifying the actual DNA sequence. Epigenetic alterations may be triggered in aquaculture by dietary inputs, including particular nutrients or bioactive chemicals. These changes have the potential to impact heritable characteristics that are inherited by future generations. Nutrients have the ability to modify the configuration of DNA or the histone proteins in chromatin, resulting in persistent changes in gene expression. Epigenetic effects are caused by the process of DNA methylation or the modification of histones by methylation, acetylation, or biotinylation. These mechanisms may act alone or in combination (Mohanty et al., 2020). Epigenetic modifications may induce alterations in gene expression that last for the whole lifespan of the organism and may even be inherited by future generations. By incorporating nutrigenomics into aquaculture methods, a more refined and sophisticated method of designing feeding regimens becomes possible. Aquaculturists get valuable knowledge about the distinct dietary requirements of various species and people within a community by decoding the genetic and epigenetic information. The accuracy of this approach allows for the development of feeds that take into account the genetic variation and epigenetic adaptability of aquatic creatures, thus enhancing their growth, well-being, and general viability in aquaculture activities. As we explore the intricate relationship between genetic and epigenetic elements in nutrigenomics, it becomes clear that there is great potential for significant breakthroughs in aquaculture techniques. These developments may lead to a more efficient, sustainable, and ecologically conscious sector. SNPs, which are common genetic variations, have the ability to modify the expression or activity of genes. As a result, they may impact nutrient metabolism and the body's need for nutrients. SNPs may lead to either modifications in gene expression or modifications in the gene product, resulting in alterations to the structure and function of proteins. These SNPs remain genetically associated with each other and are passed down over several generations. Several frequently occurring SNPs have been identified to impact nutritional needs (Figure 5.1).

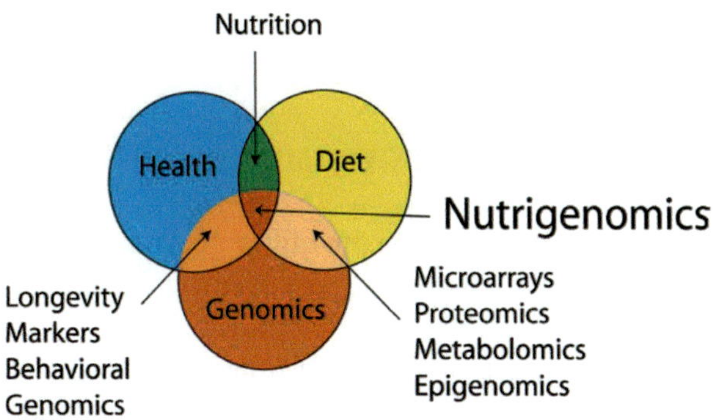

FIGURE 5.1 Venn diagram illustration of nutrigenomics as the intersection between health, diet, and genomics.

Source: Reprinted with permission from Ordovas (2008). Copyright 2008. North American Association for the Study of Obesity (NAASO)

5.4 TOOLS IN NUTRIGENOMICS

Currently, the majority of nutrigenomics research in aquaculture uses the transcriptomic method, which looks at the transcription products, or mRNA, which provide information on how nutrients affect a gene. Reverse-transcription polymerase chain reaction (PCR) is the basis for transcriptomic investigations. After extraction, the whole mRNA is reverse transcribed to cDNA. This cDNA is suitable for a variety of PCR processes. The two methods most frequently employed to assess the differential transcriptome abundance in the various treatments after a nutritional interference are semiquantitative PCR and real-time PCR. An emerging method in fish nutritionomics is the DNA microarray, which provides the organism's total genomic response to the nutrient or diet. However, because it needs the genomic sequences, its application in fish nutrition is restricted. Recombinant DNA can be inserted into a vector—a carrier of DNA molecules that will replicate recombinant DNA fragments in host organisms—using a series of techniques called molecular cloning.

The standard four-step cloning methodology is as follows:

1. Target DNA fragment isolation.
2. Inserts are ligated into the proper cloning vector to produce recombinant molecules.
3. Recombinant plasmids are being transformed into bacteria or another appropriate host for proliferation.
4. The technique is also known as qPCR because it is used to detect the quantity of a target sequence in real time by screening hosts carrying the appropriate recombinant plasmid.

5.5 APPLICATION OF NUTRIGENOMICS IN AQUACULTURE

Nutrigenomics is used in aquaculture to utilize genetic data for the purpose of optimizing nutrition and improving the overall performance of aquatic creatures. This strategy seeks to enhance growth rates, illness resistance, and nutrient utilization efficiency by customizing dietary formulations according to unique genetic profiles. Nutrigenomics facilitates a targeted approach, enabling the creation of customized feed formulas that specifically target the distinct genetic differences present in various animals and persons. This application has the capacity to completely transform aquaculture methods by promoting sustainability and efficiency, while also reducing environmental harm. Additionally, it provides a scientific basis for improving targeted nutritional treatments in the development of aquatic species. It enhances the use of nutrients and improves the efficiency of feed by customizing diets based on individual genetic differences. This precise method improves the absorption of nutrients, reduces waste, and stimulates efficient development, hence supporting sustainable aquaculture methods and avoiding environmental harm.

Some other ways in which nutrigenomics research may be used in aquaculture are as follows:

1. For assessing how the organisms react to the nutrients.
2. Diet development by enhancing a given species' ability to utilize dietary nutrients.
3. Promote in-depth knowledge of a cell's nutritional response.
4. Enhancing fundamental husbandry techniques.
5. Determine the elements in charge of metabolism.
6. Encourage understanding of the organs and tissue specialization for various nutrient utilization.
7. Knowing how dietary variables affect fundamental metabolic regulation.
8. Research on nutrigenomic analysis in nutrient transport.

5.6 CHALLENGES AND FUTURE PERSPECTIVES

The challenges in nutrigenomics for aquaculture include deciphering intricate gene–nutrient relationships, comprehending species-specific reactions, and incorporating extensive genomic data. In order to comprehensively comprehend the effects of aquaculture feeds on fish physiology, it is necessary to adopt a distinct methodology that focuses on identifying the molecular and cellular mechanisms that control the reactions to various diets. Several issues arise in the context of nutrigenomic and nutrigenetic data, including the need for quality control, standardization, and the acquisition and storage of data. The "omic" techniques generate large volumes of data quickly. Significant advancements have been achieved thus far, and by utilizing novel analytical techniques on the existing data, it has been demonstrated that it is not only feasible to compile inventories of gene products and metabolites that undergo alterations in specific circumstances but also to acquire a deeper understanding of the underlying biological processes.

This emerging omics technique has significant promise for studying the intricate correlation between fish nutrition and gene expression. Future prospects include the optimization of individualized dietary regimens, the use of modern technology such as clustered regularly interspaced short palindromic repeats (CRISPR) for genetic augmentation, and the resolution of ethical concerns. To fully harness the transformative power of nutrigenomics in changing aquaculture methods toward sustainability and efficiency, it is imperative to get a thorough understanding of the dynamics of nutrigenomics and promote cooperation within the fields of genetics, nutrition, and aquaculture. The future of nutrigenomics in aquaculture will be influenced by ongoing research, technical advancements, and the implementation of ethical frameworks.

5.7 ETHICAL AND REGULATORY CONSIDERATIONS

The ethical and regulatory aspects of nutrigenomics in aquaculture require the careful management of possible environmental consequences, the protection of animal well-being, and the preservation of genetic authenticity. Ensuring a harmonious equilibrium between scientific progress and ethical principles is of utmost importance. Enforcing clear protocols for genetic alterations, upholding innate tendencies, and reducing unexpected repercussions are moral obligations. Regulatory frameworks need to adapt to effectively govern the ever-changing field of genetic therapies, with a focus on promoting ethical research and industrial practices. In order to maintain sustainability, animal welfare, and public confidence in the proper use of nutrigenomics in aquaculture, it is crucial to implement proactive ethical and regulatory measures.

5.8 CONCLUSION

Nutrigenomics offers a novel approach to studying nutrition, focusing on the interaction between food and the genetic code, as well as the resulting physiological responses and observable traits. Nutrigenomics has great promise for identifying genes that play a crucial role as targets for dietary interventions. Furthermore, it should play a crucial role in clarifying nutrient signaling pathways that might potentially contribute to certain diet-related disorders. The integration of nutrition sciences with communication and marketing has led to the creation of individualized nutritional advising rooted in nutrigenomics. Scientists and experts in the field of human health sciences may continuously contribute to nutrigenomics by conducting research and developing new tools that can improve the quality of life and promote a healthy diet for the public.

KEYWORDS

- **nutrigenomics**
- **feed**
- **gene expression**
- **growth performance**
- **disease resistance**

REFERENCES

Arshad, Z., Bains, V. K., Jhingran, R., Madan, R., & Srivastava, R. (2017). Nutrigenomics: An overview. *Asian Journal of Oral Health and Allied Sciences*, *7*, 52–61.

Cozzolino, S. M. F., & Cominetti, C. (2013). *Biochemical and Physiological Bases of Nutrition in Different Stages of Life in Health and Disease*. Monole: Sao Paulo, Brazil.

Craig, S. R., Helfrich, L. A., Kuhn, D., & Schwarz, M. H. (2017). *Understanding Fish Nutrition, Feeds, and Feeding*, Virginia Cooperative Extension.

DeBusk, R. (2009). Diet-related disease, nutritional genomics, and food and nutrition professionals. *Journal of the American Dietetic Association*, *3*(109), 410–413.

Drayna, D. (2005). Human taste genetics. *Annual Review of Genomics and Human Genetics*, *6*, 217–235.

Fuhrman, J. A. (1999). Marine viruses and their biogeochemical and ecological effects. *Nature*, *399*(6736), 541–548.

Gardy, J. L., & Loman, N. J. (2018). Towards a genomics-informed, real-time, global pathogen surveillance system. *Nature Reviews Genetics*, *19*(1), 9–20.

Ghoshal, S., Pasham, S., Odom, D. P., Furr, H. C., & McGrane, M. M. (2003). Vitamin A depletion is associated with low phosphoenolpyruvate carboxykinase mRNA levels during late fetal development and at birth in mice. *The Journal of Nutrition*, *133*(7), 2131–2136.

Holliday, R., & Grigg, G. W. (1993). DNA methylation and mutation. *Mutation Research/Fundamental and Molecular Mechanisms of Mutagenesis*, *285*(1), 61–67.

Lende, S. R., Mahida, P. J., Shrivastava, V., Yusufzai, S. I., & Jha, A. K. (2014) Nutrigenomics: The new era of aquaculture nutrition. *Indian Farmer*, *1*(Spl. I), 320–328.

Martin, S. A., & Król, E. (2017). Nutrigenomics and immune function in fish: New insights from omics technologies. *Developmental & Comparative Immunology*, *75*, 86–98.

Menon, B., Harinarayan, C. V., Raj, M. N., Vemuri, S., Himabindu, G., & Afsana, T. K. (2010). Prevalence of low dietary calcium intake in patients with epilepsy: A study from South India. *Neurology India*, *58*(2), 209.

Mohanty, B. P., Ganguly, S., Mahanty, A., Mitra, T., & Mohanty, S. (2020). Nutrigenomics and fish. *CABI Reviews*, *15*(048), 1–19.

Ordovas, J. M. (2008). Genotype–phenotype associations: Modulation by diet and obesity. *Obesity*, *16*(S3), S40–S46.

Ruden, D. M., De Luca, M., Garfinkel, M. D., Bynum, K. L., & Lu, X. (2005). Drosophila nutrigenomics can provide clues to human gene-nutrient interactions. *The Annual Review of Nutrition*, *25*, 499–522.

CHAPTER 6

Functional Feed Additives in Aquaculture: Enhancing Health, Growth, and Sustainability

RAJKUMAR DABHI[1*], KRINA PATEL[2], YAGNESH MOTIVARASH[3], KAILASH VADHER[2], KIRAN JORA[2], and PINAK BAMANIYA[4]

[1]*Department of Aquaculture, College of Fisheries Science, Kamdhenu University, Navsari*

[2]*Department of Aquaculture, College of Fisheries Science, Kamdhenu University, Veraval, Gujarat, India*

[3]*Department of Fisheries Resource Management, College of Fisheries Science, Kamdhenu University, Veraval, Gujarat, India*

[4]*Department of Aquatic Environment Management, College of Fisheries Science, Kamdhenu University, Veraval, Gujarat, India*

*Corresponding author

ABSTRACT

The global fisheries and aquaculture sector are vital for food security and nutrition, especially with the world's population expected to reach 9 billion by 2050. As the demand for food increases, aquaculture is becoming the fastest-growing food production system, offering food, jobs, and income to millions. To meet this growing demand, new and improved aquaculture practices are needed. One such innovation is the use of functional feed additives. These are special ingredients added to fish and shrimp feed that not only meet nutritional needs but also improve health, growth, and disease resistance. They help aquatic animals cope with stress and fight infections by strengthening their immune systems and reducing oxidative stress.

Advances in Freshwater Aquaculture: Technology for Sustainable Development. Tarang Kumar Shah, Madhu Sharma, Mohd. Ashraf Rather, & Avdhesh Kumar (Eds.)

DOI: 10.1201/9781779640857-6

Functional additives include amino acids, vitamins, minerals, enzymes, probiotics, algae, mushrooms, yeast, phytogenic compounds, and organic acids. These improve digestion, growth, and survival rates. When used in combination, they can have even stronger effects, making aquaculture more productive and sustainable. However, to get the best results, it is important to understand how these additives interact with each other, the feed, and the animals' biology. This chapter explains the types and benefits of functional feed additives and their role in sustainable aquaculture development.

6.1 INTRODUCTION

Aquaculture has emerged as a vital contributor to global food production to meet the increasing demand for seafood. As this industry continues to grow, there is a corresponding need to enhance the efficiency and sustainability of production systems (Dewali et al., 2023). The main aspect influencing the success of aquaculture operations is the health and performance of the cultured organisms (Bohnes et al., 2022). In this context, the role of functional feed additives has gained importance as a strategic tool in enhancing aquaculture practices. Functional feed additives are specialized compounds incorporated in the diets of aquatic organisms to effect specific physiological responses that promote growth, health, and overall performance (Pandey et al., 2019).

The increasing intensification of aquaculture operations also raises environmental concerns, which demand more responsible and sustainable practices (Boyd et al., 2020). As aquaculture systems become more intensive, the risk of diseases spreading rises. Functional feed additives provide a better approach to fight against specific challenges in aquaculture systems, contributing to the overall well-being of cultured organisms and the economic viability of the industry.

Excessive use of antibiotics, hormones, and other synthetic chemicals in aquaculture to control diseases and improve fish production has resulted in the creation of drug-resistant bacteria, immune system inhibition, and accumulation in animal tissues and the production of toxic compounds harmful to the environment and human health (Esiobu et al., 2002; Ringø et al., 2010; Gomez-Zavaglia et al., 2019) and suppress immunity in the host (Panigrahi and Azad, 2007). The use of such antibiotics in aquaculture has led to the evolution and spread of several resistant strains of pathogens like *Aeromonas* sp., *Escherichia tarda*, *Escherichia coli*, *Vibrio vulnificus*, *Vibrio parahaemolyticus*, *Vibrio cholerae*, and many more (Brogden et al.,

2014; Allameh et al., 2016). Thus, their use has been criticized all over the world (Baruah et al., 2008).

There is a need to find sustainable alternative strategies to improve disease resistance in cultured fish (Dawood et al., 2018). Natural immunostimulants, such as prebiotics, synbiotics, probiotics, nutritional supplements, complex carbohydrates, hormones, herbs, and cytokines, may be helpful techniques for promoting development, stimulating the immune response, and controlling disease in aquatic animals (Al-Saif et al., 2014; De Jesus Raposo et al., 2015; Dawood et al., 2022).

Over the past decade, the application of functional feed additives has been practiced, improving nutrient digestibility and digestive enzyme activities in the aquaculture fish and shellfish (Song et al., 2014; Van Hai, 2015; Hoseinifar et al., 2016; Ringø and Song, 2016). Different types of functional feed additives, such as immunostimulants, prebiotics, probiotics, and other bioactive compounds, play a crucial role in enhancing the immune systems of aquatic organisms. This minimizes the need for antibiotics, and it pushes global efforts toward antibiotic-free and environmentally sustainable aquaculture practices. This chapter summarizes and evaluates the broader knowledge about the functional feed additives.

6.2 DEFINITION

Functional feed: The feed that serves a specific purpose beyond the basic nutritional requirement to promote health and stress resistance of the animals.

Functional additives: The additives that serve a specific purpose beyond the basic nutritional requirement to promote health and stress resistance of the animals.

6.3 IMPORTANCE OF FUNCTIONAL FEED ADDITIVES IN AQUACULTURE

Functional feed additives hold immense importance in the field of aquaculture, contributing significantly to the success, sustainability, and efficiency of aquatic farming operations. Several key factors underscore the importance of functional feed additives in aquaculture:

Nutritional optimization: There are specific nutritional requirements of every aquatic species for growth, reproduction, and overall health.

The use of functional feed additives supports the nutritional content of aquafeed to meet the requirements and ensure sustainable development.

Enhanced growth and performance: Many functional feed additives, such as enzymes and amino acids, enhance the digestibility and absorption of nutrients. This resulted to improved growth rates, better feed conversion ratios, and overall enhanced performance of the cultured species.

Disease prevention and health promotion: Some additives play a significant role in preventing diseases and improve the overall health and resilience of the population, such as Probiotics, prebiotics, and immunostimulants. These additives also contribute to a healthy gut microbiota and immune system in aquatic organisms.

Environmental sustainability: Functional feed additives can reduce the environmental effects of aquaculture by increasing the efficiency of nutrient consumption. This involves reducing waste output and nutrient runoff, both of which are crucial for reducing the ecological impact of aquaculture operations.

Stress reduction: For fish and other aquatic organisms, aquaculture environments can be stressful due to factors like crowding, handling, and changes in water quality. Functional additives, such as stress-reducing compounds, help mitigate the negative effects of stress, leading to healthier and more resilient populations.

Economic benefits: The use of functional additives can enhance growth rates and improve feed conversion efficiency and contribute to economic benefits for aquaculture producers. Functional feed additives help maximize production yields, reduce production costs, and improve the overall profitability of aquaculture operations.

Customization for species-specific needs: The nutritional requirements of different aquatic species are unique. Different functional feed additives allow for the customization of aquafeed formulations to address the specific needs of each species, ensuring optimal growth and health.

Substitution of expensive ingredients: The partial replacement of expensive ingredients, such as fish oil or meal, with cheaper and sustainable substitutes is possible through specific functional feed additions. Aquaculture becomes more environmentally sustainable as a result in reducing the reliance on marine resources.

Compliance with regulations: As the aquaculture sector develops, sustainable and responsible methods are becoming more and more important. Functional feed additives can help aquaculture companies comply with regulatory requirements to antibiotic use, environmental effects, and general animal welfare.

6.4 DIFFERENT GROUPS OF FEED ADDITIVES USED IN AQUACULTURE FEED

Functional feed additives in aquaculture can be categorized into various groups, each serving specific purposes to enhance the overall health, growth, and performance of aquatic species. Here are some detailed groups of functional feed additives.

6.4.1 AMINO ACIDS

Amino acids are known as the building blocks of protein. Based on dietary requirements for growth, amino acids were usually classified as essential or nonessential for fish. Some are conditionally essential amino acids that must be provided from the diet when rates of utilization are greater than rates of synthesis. There are many studies that show amino acids regulate metabolic pathways for the maintenance, growth, reproduction, and immune response (Figure 6.1) in aquatic and terrestrial animals, which makes them "functional amino acids" (Li et al., 2009).

According to recent research, some amino acids and their metabolites play a significant role in regulating vital metabolic pathways that fish require for survival, development, immunity, behavior, larval metamorphosis, feeding, nutrient utilization, reproduction, and resistance to pathogens and environmental stressors (Li et al., 2009) (Table 6.1).

6.4.2 PHYTOGENIC OR PHYTOBIOTIC COMPOUNDS

Phytobiotics are a broad class of naturally occurring, plant-based, bioactivated compounds that are added to animal feed to improve animal productivity, disease resistance, and the quality of animal-derived food products. They are also referred to as phytogenics, phytochemicals, or botanicals (Hashemzadeh-Cigari et al., 2014).

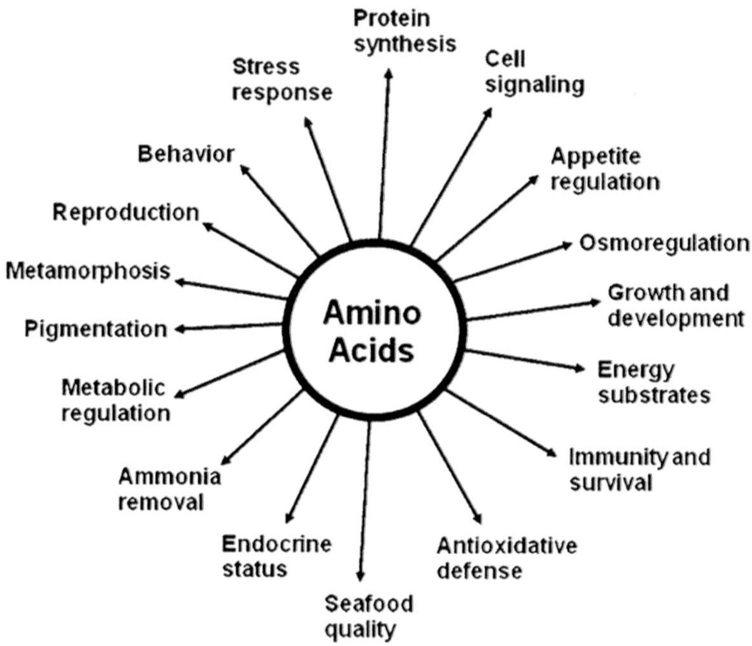

FIGURE 6.1 Roles of amino acids.

Phytogenic or phytobiotic compounds are used nowadays in aquaculture feed industries as an additive due to their eco-friendly nature. They are helpful for growth enhancement and antimicrobial performance in aquatic organisms. There are many synthetic antibiotic compounds available in the market that are used in aquaculture, but they are nowadays restricted or banned for use due to the potential to develop antibiotic-resistant pathogenic bacteria. The bioactive compounds, such as alkaloids, tannins, saponins, glycosides, terpenoids, phenolics, flavonoids, essential oils, or steroids that are found in phytobiotics, which serves in various purposes, including enhancing immunity, promoting growth, acting as bio-appetizers, stimulating reproduction, and functioning as bio-preservatives. Therefore, phytobiotics are emerging as a beneficial supplement in aquaculture (Figure 6.2), acting as an alternative to antibiotics and other chemical

TABLE 6.1 Effect of Different Amino Acids in the Diet for Various Aquatic Organisms

Amino Acids	Effect	Reference
Arginine	Dietary arginine levels had a significant effect on the growth and immune responses of golden pompano	Lin et al. (2015)
Histidine	Enhanced antioxidant capacity and protein synthesis for juvenile swimming crabs	Tian et al. (2023)
Isoleucine	Growth, carcass composition, hematological status, serum protein and antioxidant enzyme activity of fingerling *Channa punctatus* were significantly affected by the levels of isoleucine in the diet	Sharf and Khan (2020)
Leucine	Increased growth performance, nonspecific immune responses, and gut morphology of golden pompano	Tan et al. (2016)
Lysine	Improved fish growth, hematobiochemical parameters, immune response, and digestive ability, and also upregulate the antioxidant status in *Oncorhynchus mykiss*	Ahmed and Ahmad (2021)
Methionine	Beneficial effect on fish feed efficiency, growth, body protein content, and digestive enzyme activity of catfish, *Pangasianodon hypophthalmus* fingerlings	Rachmawati et al. (2023)
Phenylalanine	Effectively improved the growth, digestion, absorption capacity, antioxidant capacity, and intestinal health of *Oncorhynchus mykiss*	Zhang et al. (2023)
Cystine and tyrosine	Significant effect on the melanin content, tyrosinase activity, tyrosinase gene, and gene mRNA expressions in the skin of Malaysia red tilapia	Wang et al. (2018)
Glutamate and glycine	Improved the digestibility of protein, lipid, and most of the amino acids and fatty acids in rainbow trout	Belghit et al. (2023)
Hydroxyproline	Enhances growth in Atlantic salmon fed diets high in plant protein sources in *Salmon salar*	Aksnes et al. (2008)
Proline	Improve anti-oxidative capacity, immune response, NH_3 stress tolerance of *Litopenaeus vannamei*	Xie et al. (2015)
Taurine	Improved growth performance, increased capacities of digestion, antioxidant capacity, regulated lipid metabolism, and modulated TOR gene expression in black carp, *Mylopharyngodon piceus*	Zhang et al. (2018)
Alanine and glycine	Improve the growth performance, liver function, and protein utilization of *Ctenopharyngodon idella*	Jin et al. (2016)

TABLE 6.1 *(Continued)*

Amino Acids	Effect	Reference
Aspartate	Improve the ability to fight against *Aeromonas hydrophila* infection in *Cyprinus carpio*	Zhao et al. (2022)
Valine	Enhanced the growth performance, antioxidant capacity, and mitigates *Aeromonas veronii*-induced intestinal injury in largemouth bass, *Micropterus punctulatus*	Cao et al. (2024)
Tryptophan	Improved fish growth, digestive and absorptive function as well as protein synthesis in *Cyprinus carpio*	Tang et al. (2013)
Threonine	Enhanced growth and increased capacities of digestive and absorptive, fish protein synthesis ability, improved growth and development of hepatopancreas and intestine in Jian carp (*Cyprinus carpio*)	Feng et al. (2013)
Phenylalanine and tyrosine	Increases aquaculture productivity of *Heteropneustes fossilis* fry while preventing nitrogenous waste and environmental damage caused by imbalanced dietary essential nutrients	Sayed and Ahmed (2022)

compounds (Bhanja et al., 2023). Over the past 20 years, there has been an increase in the use of phytobiotics in animal feed (Mohammadi Gheisar and Kim, 2018).

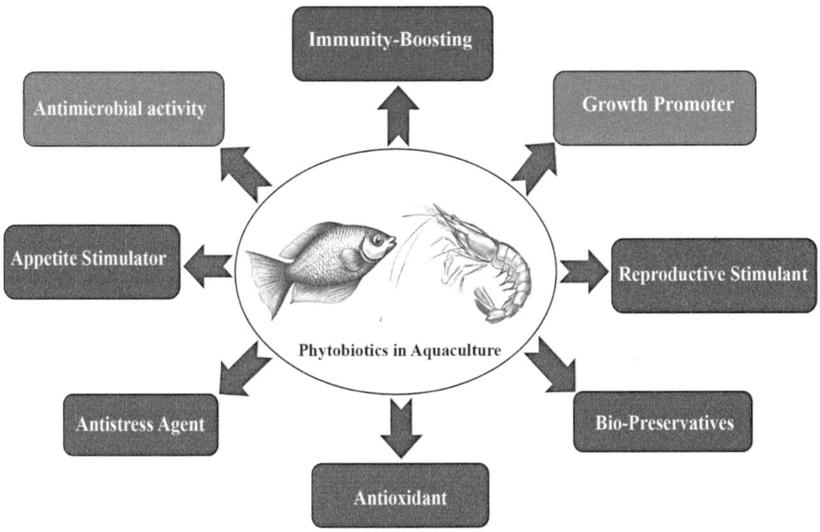

FIGURE 6.2 Roles of phytogenic or phytobiotic compounds in aquaculture.

6.4.3 MICROALGAE

Microalgae are ubiquitous photosynthetic microorganisms that are found in aquatic environments and are important as primary producers in aquatic ecosystems (Brown and Blackburn, 2013). Microalgae include eukaryotic cells such as Chlorophyceae (green algae), Bacillariophyceae (diatoms algae), and Chrysophyceae (golden algae). The term microalgae is also used for prokaryotic microorganisms. Microalgae play a crucial role as a primary food source for larval bivalves, shrimps, and fish, owing to their nutritional richness and suitable-sized cells. Especially, bivalves depend mostly on microalgae as a direct source of nutrition throughout their entire life cycle (Ma and Hu, 2023).

Microalgae such as *Chlorella* sp., *Spirulina* sp., *Tetraselmis* sp., *Isochrysis* sp., *Pavlova* sp., *Phaeodactylum* sp., *Chaetoceros* sp., *Skeletonema* sp., *Thalassiosira* sp., *Arthrospira platensis*, *Pavlova* sp., *Dunaliella* sp., *Haematoccous* sp., and *Nannochloropsis* sp. are mainly used in aquafeeds due to their superior nutritional content and appropriateness for aquaculture (Spolaore et al., 2006; Khatoon et al., 2010).

The content of protein, lipids, and carbohydrates in microalgae are 12–65%, 2–23%, and 5–26% (dry-basis), respectively (Becker, 2013). Also, microalgae are rich in biologically active chemicals such as omega-3 long-chain polyunsaturated fatty acids (PUFAs), pigments, polysaccharides, amino acids, proteins, fibers, antioxidant pigments, and vitamins and minerals. This makes them well suited for use as sustainable additives in aquafeed (Chen et al., 2021; Siddik et al., 2023). Therefore, microalgae are important sources for aquaculture species and can also be used as potential protein/lipid sources and additives for aquafeeds (Yarnold et al., 2019). The utilization of microalgae biomass has been proposed as a valuable and sustainable feed for fish and shrimp in aquaculture (Camacho-Rodríguez et al., 2018; Ahmad et al., 2023).

6.4.4 SEAWEEDS

Seaweeds are generally inhabited in the littoral zone and can be of many different shapes, sizes, colors, and compositions. They include brown algae (Phaeophyceae), red algae (Rhodophyceae), and green algae (Chlorophyceae). The composition of seaweeds can significantly differ depending on the species, habitat, collection time, and environmental conditions such as water temperature, light levels, and nutrient content (Makkar et al., 2016).

Seaweeds are a vital source of beneficial metabolites such as pigments, carotenoids, phlorotannins, PUFAs, agar, alginate, and carrageenan. In addition, they also contain essential minerals like iodine, zinc, sodium, calcium, manganese, iron, and selenium. Due to this rich composition, seaweeds are recognized as a natural reservoir of additives that have the potential to replace antibiotics in various animal feeds (Morais et al., 2020).

Seaweeds generally contain low levels of lipids; however, a significant portion of these lipids comprises PUFAs, with a notable presence of 20:5 n-3 eicosapentaenoic acid and 20:4 n-5 arachidonic acid, particularly in brown and red algae (Mišurcová, 2011). These properties make them highly attractive for utilization as ingredients in aquafeed or for extracting PUFAs as supplements. The protein content can reach up to 50% in dry weight for certain *Rhodophyta* species, and it changes based on the genus, species, season, and environmental conditions of the algae (Wan et al., 2019).

Moreover, macroalgae can be an excellent source of bioactive compounds. They produce a wide range of secondary metabolites, including free amino acids, peptides, lectins, phycobiliproteins, carotenoids, phenolics, vitamins, and functional carbohydrates, which are generally linked to their adverse environment and physical stressors (Wan et al., 2019).

Seaweeds extract and functional compounds contain sulfated polysaccharides, such as alginate, fucoidan, carrageenan, laminarin, ulvan, galactan, and agar. These compounds show various physiological activities, such as antiviral, antibacterial, anti-inflammatory, anthelmintic, antioxidant, immunomodulatory, and anticoagulant properties (Safavi et al., 2019).

6.4.5 MUSHROOMS

Mushrooms are rich sources of prebiotic compounds as hemicellulose, chitin, α- and β-glucans, mannans, xylans, and galactans. They contain several types of polysaccharides, all of which are potential prebiotics (Cheung, 1996; Kalač, 2009; Mohan et al., 2022). Polysaccharides derived from mushrooms play a crucial role as prebiotics and exhibit various biological properties, including antitumor, antimicrobial, antioxidant, antiviral, and immunomodulatory activities (Lindequist et al., 2005; Thatoi and Singdevsachan, 2014). In recent times, mushrooms have an increased attention as potential natural prebiotics in aquaculture (Lindequist et al., 2005; Aida et al., 2009; Van Doan et al., 2016; Zou et al., 2016).

- *Pleurotus*, a mushroom genus, which includes several edible mushroom species, also ranks third in the commercial production of edible mushrooms in the world. These mushrooms have been studied to have antitumor, antiviral, immunostimulants, antioxidant, antimicrobial, antimitogenic, and antiproliferative activities (Sartori et al., 2015).
- *Lentinula edodes* is commonly known as Shitake and widely commercialized for both culinary and medicinal purposes. This mushroom is abundant in bioactive compounds that have proven efficacy in addressing conditions such as hypertension, hyperlipidemia, cardiovascular irregularities, compromised immunity, hepatic disorders, and cancer (Van Doan et al., 2019). A polysaccharide called lentinan,

extracted from *L. edodes*, has shown immunomodulation properties in cultured fish (Djordjevic et al., 2009; Ren et al., 2019), while extract or whole mycelium also improved growth performance (Chitsaz et al., 2018; Pascual et al., 2018).

- *Coriolus versicolor*, a renowned mushroom employed in Asian traditional medicine, is utilized for the treatment of cancer and diseases related to immune deficiency (Van Doan et al., 2019). Its primary attributes include antioxidant, antibacterial, antitumor, immunomodulatory, and anticancer properties. Limited studies on its effects in cultured fish indicate a broad immunomodulatory impact and, notably, an enhancement in the growth performance of *Paralichthys olivaceus* (Mohan et al., 2022).

Medicinal mushrooms like *Cordyceps sinensis*, *Schizophyllum commune*, and *Inonotus obliquus* are undergoing tests as additives in fish feed to a limited extent. These mushrooms demonstrate positive effects on various aspects such as immunological parameters, growth, and stress (Van Doan et al., 2019).

6.4.6 YEASTS

Yeast, also known as mold or fungus, is a single-celled eukaryote that has membrane-bound organelles such as the mitochondria, nucleus, and endomembrane system (Behera et al., 2019). They exhibit budding or fission reproduction and are commonly found in various natural habitats, including soil, freshwater, and seawater (Navarrete and Tovar-Ramírez, 2014).

Yeasts are a vital source of protein and B-complex vitamins. They have been effectively incorporated into fish diets as an additional source of protein (Liao, 1977). Additionally, because they are easily manufactured on an industrial scale from a variety of carbon-rich substrate by-products, they are thought to be an economical nutritional supplement (Lee and Kim, 2001). Yeasts have been shown to promote immunological and antioxidant systems, as well as growth, survival, and maturity in fish and crustaceans. β-Glucan, mannoproteins, nucleic acids, and chitin (as a minor component) are the primary components of yeast's immune-stimulatory properties (Meena et al., 2013). Mannan-oligosaccharide present in yeast removes the bacteria from the gut, enhances growth performance, improves feeding efficiency, and increases gut absorption efficiency (Dimitroglou et al., 2010).

Crude protein content varies in yeast products; it can be as high as 67.8% in *Candida lipolytica* and as low as 38.8% in *Candida boidinii* (Schulz and Oslage, 1976). Because of this, the nutritional value of yeast products varies depending on their type. *Candida* sp., *Hansenula* sp., *Pichia* sp., and *Saccharomyces* sp. are some important yeast species that have been used in fish feeds (Yeh, 1963).

Additionally, it is reported that the dry matter that makes up 93–97% of dry yeast on average is made up of 40–60% crude nitrogen protein, 5% lipids, and 35–45% carbohydrates. Many mineral components are available in yeast, such as sodium, copper, zinc, calcium, manganese, iron, and selenium. Furthermore, linoleic and alpha-linolenic acids, as well as unsaturated fatty acids, are present in yeast species like *Saccharomyces cerevisiae* and can be produced by the enzyme fatty acid desaturase (Zhang et al., 2009; Bertolo et al., 2019).

Aquaculture is becoming more interested in using yeasts due to their benefits, which include the ability to produce a variety of bioactive compounds, such as glucans, glutathione, toxins, enzymes, phytase, nucleotides, and vitamins and the potential for application in bioremediation of water quality (Ernesto Ceseña et al., 2021). There are many species of yeasts such as *S. cerevisiae*, *Kluyveromyces*, *Torulaspora*, *Saccharomyces*, and *Torulopsis* are used as a source of protein in aquafeeds for the culture of larvae of shrimps and marine fishes (Glencross et al., 2020).

6.4.7 PROBIOTIC/PREBIOTIC/SYNBIOTIC

6.4.7.1 PROBIOTICS

Probiotics serve as live microbial supplements in fish feed, promoting growth by influencing the composition of gut microbial flora. They can consist of a single microorganism species or a blend of species. These microorganisms establish a presence in the gut, displacing harmful species and enabling the fish to conserve metabolic energy that would otherwise be expended in combating detrimental organisms. It is essential to incorporate probiotics into feeds postpelleting for optimal effectiveness (Pandey et al., 2019). Probiotics enhance the body's natural defense systems, inhibit harmful microorganisms, and regulate the proliferation of gut microbiota.

Probiotics increase feed conversion ratios and lower mortality rates when incorporated into diets. It has been found that probiotics can boost fish immunity and improve fish immune systems (Kumar et al., 2016).

The most common species of bacteria used as probiotics include *Bifidobacterium*, *Lactobacilli*, *Enterococcus faecium*, and spore-forming *Bacillus* spp. Some yeasts, including *Saccharomyces*, are also utilized (Pandey et al., 2019). There are many studies shows improved growth, immune response, and resistance to diseases of various species of fish and crustaceans due to the effect of probiotic bacteria (i.e., *Bacillus* spp., *Lactobacillus* sp., *Pediococcus*, *Enterococcus*, *Lactococcus*, *Aeromonas*, *Alteromonas*, *Arthrobacter*, *Bifidobacterium*, *Brochothrix*, *Clostridium*, *Kocuria*, *Microbacterium*, *Micrococcus*, *Paenibacillus*, *Phaeobacter*, *Pseudoalteromonas*, *Pseudomonas*, *Rhodococcus*, *Rhodopseudomonas*, *Rhodosporidium*, *Roseobacter*, *Shewanella*, and *Vibrio*) (Ringø et al., 2022; Van Doan et al., 2021; Yilmaz et al., 2022).

6.4.7.2 PREBIOTICS

Prebiotics are described as nondigestible food components that have a positive impact on the host by selectively promoting the growth and/or function of a specific group of bacteria in the colon. This, in turn, enhances the overall health of the host. Prebiotics serve as nourishment for bacterial species deemed beneficial for well-being and are valuable dietary supplements for regulating the development and behavior of bacterial strains in the colon (Gibson and Roberfroid, 1995; Ringø et al., 2010).

Prebiotics make a specific alteration of the gut microbiota, specifically increased numbers of *bifidobacteria* and/or *lactobacilli* cell numbers or a decrease in potentially harmful bacteria is enough reason for health promotion (Ringø et al., 2010). Prebiotics such as β-glucan fructooligosaccharide (FOS), oligofructose, inulin, xylo-oligosaccharide, and mannan oligosaccharide are frequently used in fish diets (Crittenden and Playne, 2009).

Prebiotics are fermented by bacteria in the host's gut instead of being directly digested by the host. Bacteria such as *Bifidobacterium* and *Lactobacillus* play a crucial role in producing short-chain fatty acids (SCFAs), including acetic, propionic, and butyric acids. This fermentation process results in a decrease in gut pH, contributing to both colonic and systemic health effects. The host may absorb these SCFAs, utilizing them as energy sources (Guerreiro et al., 2018). Prebiotics contribute to the improvement of diverse genetic reactions in aquatic organisms and alleviate the harmful impacts of heavy metals (Mustafa and Al-Faragi, 2021). Moreover, they serve as immunostimulants, strengthening the host's immune defenses against incoming pathogens (Dawood et al., 2018).

6.4.7.3 SYNBIOTICS

Synbiotics, which are feed supplements comprising a combination of prebiotics and probiotics, have the potential to enhance various aspects of health. Despite their known benefits in improving immune responses, there is currently a lack of extensive research on the application of synbiotics in cultured fish, and the available information remains limited. In studies involving synbiotics as feed additives, researchers have primarily focused on specific physiological parameters such as growth performance, biochemical/hematological changes, and disease resistance. Synbiotics work by promoting the survival and manipulation of live microbes in the gastrointestinal tract, facilitating the metabolism of beneficial bacteria, and activating growth processes. This, in turn, contributes to the overall improvement of the host's health condition (Cerezuela et al., 2011).

Synbiotic formulations incorporate probiotic strains such as *Lactobacilli, Bifidobacteria* spp., *Saccharomyces boulardii*, *Bacillus coagulans*, among others. Meanwhile, the primary prebiotics utilized include oligosaccharides like FOS, galactooligosaccharides, and xyloseoligosaccharide, as well as inulin and prebiotics sourced naturally from roots such as chicory, yacon, and so on (Pandey et al., 2019).

In summary, the administration of synbiotics to fish led to beneficial results in terms of their immunological responses, survival, growth, gut microbiota, increased levels of intestinal absorption, and increased health status (González-Félix et al., 2018; Hassaan et al., 2018).

6.4.8 ENZYMES

Enzymes, proteins with the ability to catalyze specific biochemical reactions, are present in both plants and animals, playing a crucial role in growth and health maintenance. Microorganisms also generate enzymes, and modern advancements have enabled the large-scale industrial production of enzymes using microorganisms. Extracting and utilizing these enzymes has become integral to various processes involved in the production of feed and natural products (Pandey et al., 2019).

Exogenous enzymes, widely acknowledged as effective feed additives in formulating diets for poultry and swine, play a crucial role in mitigating the adverse impacts of antinutritional factors. They also contribute to enhancing the digestion of dietary components and ultimately improving animal performance. Numerous microbial and fungal enzyme products are currently accessible in the market. It is noteworthy that while these products are initially designed for the livestock market, they are subsequently advocated for utilization in aquaculture feeds (Encarnação, 2016).

In contemporary aquaculture practices, various exogenous enzymes such as phytase, carbohydrase, protease, and lipase are employed in fish feeds to mitigate the adverse impacts of antinutritional factors. Their purpose is to enhance the digestion of dietary components and promote the growth of fish (Lin et al., 2007; Ebru and Cengiz, 2016). Notably, phytase, an enzyme specialized in breaking down indigestible phytate, has witnessed a growing usage in fish feed over the last two decades (Cao et al., 2007).

6.4.9 VITAMINS AND MINERALS

Vitamins play a crucial role in various metabolic and biochemical reactions within the animal body, making them significant dietary components. Vitamin premixes are concentrated blends where essential vitamins are combined with a base feed. Choline chloride is deliberately excluded from these premixes due to its observed tendency to compromise the stability of other vitamins. Typically, it is incorporated into the diet at varying levels, ranging from 0.5% to 4% (Demott and Müller-Navarra, 1997).

Essential minerals in the diet include inorganic elements like calcium, phosphorus, sodium, potassium, iron, manganese, magnesium, copper, chloride, iodine, cobalt, and zinc, deficiencies of which can lead to specific diseases. Supplementing trace elements such as copper, zinc, cobalt, iodine, and manganese in the diet has been shown to enhance growth. Calcium and phosphorus are present in bones and exoskeletons, while sodium, potassium, magnesium, and chloride play roles in osmoregulation.

Additionally, magnesium, manganese, and zinc serve as cofactors for metabolic enzymes (Kyaw et al., 2010).

6.4.10 ORGANIC ACIDS

Organic acids are a type of organic carboxylic compounds characterized by the general structural formula R–COOH. The acidity of these compounds is attributed to their carboxyl group (–COOH). They exhibit weak acidic properties as they undergo partial dissociation in water, yielding a hydrogen ion (H^+) and a carboxylate ion (–COO–). An example of such an organic acid is acetic acid (CH_3COOH) (e.g., acetic CH_2COOH) (Lim et al., 2015). The practice of incorporating organic acids to achieve dietary acidification has found widespread application in animal nutrition. Organic acids are increasingly recognized as a beneficial feed additive, showing promise in enhancing gut health and overall performance (Hassaan et al., 2014). Different types of organic acids include acetic, butyric, citric, formic, lactic, propionic, malic, and sorbic acids and their salts, which have been used as acidifiers in animal feeds (NRC, 2011).

Organic acids permeate the bacterial cell wall, interfering with its usual functions and impeding their growth (Nates, 2016). Its actions are influenced by a range of factors, including the fish species, size, age, variations in organic acids, feed management practices, and the quality of the water (Fefana, 2014). Organic acids improved growth, nutrient utilization, and disease resistance of fishes (Ng et al., 2009). Organic acids have the capacity to impact the microbial composition within the digestive tract of fish, thereby exerting a crucial role in nutrition and health to enhance the immune system (Burr et al., 2005).

6.4.11 CAROTENOIDS

Carotenoids exhibit various functional attributes, serving as growth promoters, antioxidants, scavengers of free radicals, quenchers of singlet oxygen, contributors to anti-inflammatory effects, participants in antitumor activities, and enhancers of immunostimulant responses (Wan et al., 2019).

While carotenoids possess noteworthy physiological and immunological attributes, they are currently gaining more attention as technological additives. Certain carotenoids like astaxanthin and fucoxanthin are in high demand for aquafeed production due to their potential to replace synthetic colorants widely utilized in the aquaculture industry, especially in organically certified salmon production. Various studies have focused on different fish and shellfish species, revealing the efficacy of carotenoids in enhancing skin coloration (Rajauria, 2015).

6.4.12 MYCOTOXIN BINDERS

Mycotoxins are secondary metabolites that are generated by several fungal species (Binder, 2007). Mycotoxins, which are harmful metabolites, are generated by a variety of fungi such as *Aspergillus* (Manning, 2015).

Aflatoxin B1 (produced by *Aspergillus flavus*), Ochratoxin A, and *Fusarium* mycotoxins are prominent harmful substances. When aquatic species consume a diet contaminated with these mycotoxins, it can lead to a decrease in their growth and Darwinian fitness (Crittenden and Playne, 2009; Anadón et al., 2010; Song et al., 2014).

Various mycotoxin-binding agents, also known as adsorbents, have been verified and utilized to mitigate or adsorb potential levels of different mycotoxins. Common groups of mycotoxin binders employed for this purpose include aluminum silicates, hydrated sodium, montmorillonite, bentonite, and zeolitic substances (Manning, 2015).

6.5 CONCLUSIONS

The integration of functional feed additives into aquaculture practices emerges as a keystone strategy for optimizing health, growth, and sustainability in aquatic species. The importance of functional feed additives in aquaculture lies in their ability to address multifaceted challenges, such as nutritional optimization, enhanced growth, disease prevention, environmental sustainability, stress reduction, and economic benefits.

Amino acids, phytogenic compounds, microalgae, seaweeds, mushrooms, yeasts, and various probiotics/prebiotics/synbiotics play integral roles in supporting health and performance. Enzymes, vitamins, minerals, organic acids, carotenoids, and mycotoxin binders further contribute to the formulation of well-balanced and targeted aquafeeds. As aquaculture continues to evolve to meet global demands sustainably, the versatility of functional feed additives allows for precise customization to meet the unique requirements of different aquatic species.

As research and innovation persist in this field, the continued refinement and expansion of functional feed additive applications promise a resilient, sustainable, and thriving future for aquaculture.

KEYWORDS

- **aquaculture**
- **functional feed additives**
- **food security**
- **sustainability**
- **research and development**

REFERENCES

Ahmad, A., Banat, F., Alsafar, H., & Hasan, S. W. (2023). Recent breakthroughs in integrated biomolecular and biotechnological approaches for enhanced lipid and carotenoid production from microalgae. *Phytochemistry Reviews*, *22*(4), 993–1013.

Ahmed, I., & Ahmad, I. (2021). Dietary lysine modulates growth performance, haemato-biochemical indices, non-specific immune response, intestinal enzymatic activities and antioxidant properties of rainbow trout, *Oncorhynchus mykiss* fingerlings. *Aquaculture Nutrition*, *27*, 124–139.

Aida, F. M. N. A., Shuhaimi, M., Yazid, M., & Maaruf, A. G. (2009). Mushroom as a potential source of prebiotics: A review. *Trends in Food Science & Technology*, *20*(11–12), 567–575.

Aksnes, A., Mundheim, H., Toppe, J., & Albrektsen, S. (2008). The effect of dietary hydroxyproline supplementation on salmon (*Salmo salar* L.) fed high plant protein diets. *Aquaculture*, *275*(1–4), 242–249.

Allameh, S. K., Yusoff, F. M., Ringø, E., Daud, H. M., Saad, C. R., & Ideris, A. (2016). Effects of dietary mono-and multiprobiotic strains on growth performance, gut bacteria and body composition of Javanese carp (*Puntius gonionotus*, Bleeker 1850). *Aquaculture Nutrition, 22*(2), 367–373.

Al-Saif, S. S. A. L., Abdel-Raouf, N., El-Wazanani, H. A., & Aref, I. A. (2014). Antibacterial substances from marine algae isolated from Jeddah coast of Red sea, Saudi Arabia. *Saudi Journal of Biological Sciences*, *21*(1), 57–64.

Anadón, A., Martínez-Larrañaga, M. R., Caballero, V., & Castellano, V. (2010). Assessment of prebiotics and probiotics: An overview. *Bioactive Foods in Promoting Health*, 19–41.

Baruah, S. K., Norouzitallab, P., Debnath, D., Pal, A. K., & Sahu, N. P. (2008). Organic acids as non-antiobiotic nutraceuticals in fish and prawn feed. *Aquaculture Health International*, (12), 4–6.

Becker, E. W. (2013). Microalgae for aquaculture: Nutritional aspects. In: *Handbook of Microalgal Culture: Applied Phycology and Biotechnology*, Wiley, pp. 671–691.

Behera, S. S., Ray, R. C., Das, U., Panda, S. K., & Saranraj, P. (2019). Microorganisms in fermentation. In: *Essentials in Fermentation Technology*, pp. 1–39.

Belghit, I., Philip, A. J. P., Maas, R. M., Lock, E. J., Eding, E. H., Espe, M., & Schrama, J. W. (2023). Impact of dietary glutamate and glycine on growth and nutrient utilization in rainbow trout (*Oncorhynchus mykiss*). *Aquaculture*, *568*, 739311.

Bertolo, A. P., Biz, A. P., Kempka, A. P., Rigo, E., & Cavalheiro, D. (2019). Yeast (*Saccharomyces cerevisiae*): Evaluation of cellular disruption processes, chemical composition, functional properties and digestibility. *Journal of Food Science and Technology*, *56*, 3697–3706.

Bhanja, A., Payra, P., & Mandal, B. (2023). Phytobiotics: Response to Aquaculture as Substitute of Antibiotics and other Chemical Additives. *South Asian Journal of Experimental Biology*, *13*(5), 341–355.

Binder, E. M. (2007). Managing the risk of mycotoxins in modern feed production. *Animal Feed Science and Technology*, *133*(1–2), 149–166.

Bohnes, F. A., Hauschild, M. Z., Schlundt, J., Nielsen, M., & Laurent, A. (2022). Environmental sustainability of future aquaculture production: Analysis of Singaporean and Norwegian policies. *Aquaculture*, *549*, 737717.

Boyd, C. E., D'Abramo, L. R., Glencross, B. D., Huyben, D. C., Juarez, L. M., Lockwood, G. S., McNevin, A. A., Tacon, A. G., Teletchea, F., Tomasso Jr, J. R., & Valenti, W. C. (2020). Achieving sustainable aquaculture: Historical and current perspectives and future needs and challenges. *Journal of the World Aquaculture Society*, *51*(3), 578–633.

Brogden, G., Krimmling, T., Adamek, M., Naim, H. Y., Steinhagen, D., & von Köckritz-Blickwede, M. (2014). The effect of β-glucan on formation and functionality of neutrophil extracellular traps in carp (*Cyprinus carpio* L.). *Developmental & Comparative Immunology*, *44*(2), 280–285.

Brown, M. R., & Blackburn, S. I. (2013). Live microalgae as feeds in aquaculture hatcheries. In: *Advances in Aquaculture Hatchery Technology*, pp. 117–158e.

Burr, G., Gatlin III, D., & Ricke, S. (2005). Microbial ecology of the gastrointestinal tract of fish and the potential application of prebiotics and probiotics in finfish aquaculture. *Journal of the World Aquaculture Society*, *36*(4), 425–436.

Camacho-Rodríguez, J., Macías-Sánchez, M. D., Cerón-García, M. C., Alarcón, F. J., & Molina-Grima, E. (2018). Microalgae as a potential ingredient for partial fish meal replacement in aquafeeds: Nutrient stability under different storage conditions. *Journal of Applied Phycology*, *30*, 1049–1059.

Cao, L., Wang, W., Yang, C., Yang, Y., Diana, J., Yakupitiyage, A., Luo, Z. & Li, D. (2007). Application of microbial phytase in fish feed. *Enzyme and Microbial Technology*, *40*(4), 497–507.

Cao, Q., Zhao, J., He, L., Zhang, T., Feng, L., Jiang, W., Zhao, J., Liu, H., & Jiang, J. (2024). Evaluation of the dietary L-valine on fish growth and intestinal health after infection with *Aeromonas veronii*. *Aquaculture*, *580*, 740294.

Cerezuela, R., Meseguer, J., & Esteban, M. A. (2011). Current knowledge in synbiotic use for fish aquaculture: A review. *Journal of Aquaculture Research & Development S*, *1*, 1–7.

Chen, F., Leng, Y., Lu, Q., & Zhou, W. (2021). The application of microalgae biomass and bio-products as aquafeed for aquaculture. *Algal Research*, *60*, 102541.

Cheung, P. C. K. (1996). Dietary fiber content and composition of some cultivated edible mushroom fruiting bodies and mycelia. *Journal of Agricultural and Food Chemistry*, *44*(2), 468–471.

Chitsaz, H., Akrami, R., & Ahmadi, Z. (2018). Effects of mushroom (*Lentinula edodes*) extract on growth performance, immune response and hemato-biochemical parameters of great sturgeon juvenile (*Huso huso* Linnaeus, 1754). *Sustainable Aquaculture and Health Management Journal*, *4*(1), 29–48.

Crittenden, R., & Playne, M. J. (2009). Prebiotics. In: *Handbook of Probiotics and Prebiotics*. John Wiley & Sons, Inc.: New Jersey, pp. 533–562.

Dawood, M. A., El Basuini, M. F., Yilmaz, S., Abdel-Latif, H. M., Alagawany, M., Kari, Z. A., Abdul Razab, M. K. A., Hamid, N. K. A., Moonmanee, T., & Van Doan, H. (2022). Exploring the roles of dietary herbal essential oils in aquaculture: A review. *Animals*, *12*(7), 823.

Dawood, M. A., Koshio, S., & Esteban, M. Á. (2018). Beneficial roles of feed additives as immunostimulants in aquaculture: A review. *Reviews in Aquaculture*, *10*(4), 950–974.

De Jesus Raposo, M. F., De Morais, A. M. B., & De Morais, R. M. S. C. (2015). Marine polysaccharides from algae with potential biomedical applications. *Marine Drugs*, *13*(5), 2967–3028.

Demott, W., & Müller-Navarra, D. Ö. R. T. H. E. (1997). The importance of highly unsaturated fatty acids in zooplankton nutrition: Evidence from experiments with Daphnia, a cyanobacterium and lipid emulsions. *Freshwater Biology*, *38*(3), 649–664.

Dewali, S., Sharma, N., Melkani, D., Arya, M., Kathayat, N., Panda, A. K., & Bisht, S. S. (2023). Aquaculture: Contributions to global food security. In: *Emerging Solutions in Sustainable Food and Nutrition Security*. Springer International Publishing: Cham, pp. 123–139.

Dimitroglou, A., Merrifield, D. L., Spring, P., Sweetman, J., Moate, R., & Davies, S. J. (2010). Effects of mannan oligosaccharide (MOS) supplementation on growth performance, feed utilisation, intestinal histology and gut microbiota of gilthead sea bream (*Sparus aurata*). *Aquaculture*, *300*(1–4), 182–188.

Djordjevic, B., Škugor, S., Jørgensen, S. M., Øverland, M., Mydland, L. T., & Krasnov, A. (2009). Modulation of splenic immune responses to bacterial lipopolysaccharide in rainbow trout (*Oncorhynchus mykiss*) fed lentinan, a beta-glucan from mushroom *Lentinula edodes*. *Fish & Shellfish Immunology*, *26*(2), 201–209.

Ebru, Y., & Cengiz, K. (2016). Feed additives in aquafeeds. *Lucrări Științifice-Universitatea de Științe Agricole și Medicină Veterinară, Seria Zootehnie*, *66*, 155–160.

Encarnação, P. (2016). Functional feed additives in aquaculture feeds. In: *Aquafeed Formulation*. Academic Press, pp. 217–237.

Ernesto Ceseña, C., Vega-Villasante, F., Aguirre-Guzman, G., Luna-Gonzalez, A., & Campa-Cordova, A. (2021). Update on the use of yeast in shrimp aquaculture: A minireview. *International Aquatic Research*, *13*(1), 1–16.

Esiobu, N., Armenta, L., & Ike, J. (2002). Antibiotic resistance in soil and water environments. *International Journal of Environmental Health Research*, *12*(2), 133–144.

Fefana. (2014). *Organic Acids in Animal Nutrition*. Fefana Publication: Brussels, p. 97.

Feng, L., Peng, Y., Wu, P., Hu, K., Jiang, W. D., Liu, Y., Jiang, J., Li, S. H., & Zhou, X. Q. (2013). Threonine affects intestinal function, protein synthesis and gene expression of TOR in Jian carp (*Cyprinus carpio* var. Jian). *PLoS One*, *8*(7), e69974.

Gibson, G. R., & Roberfroid, M. B. (1995). Dietary modulation of the human colonic microbiota: Introducing the concept of prebiotics. *The Journal of Nutrition*, *125*(6), 1401–1412.

Glencross, B. D., Huyben, D., & Schrama, J. W. (2020). The application of single-cell ingredients in aquaculture feeds—a review. *Fishes*, *5*(3), 22.

Gomez-Zavaglia, A., Prieto Lage, M. A., Jimenez-Lopez, C., Mejuto, J. C., & Simal-Gandara, J. (2019). The potential of seaweeds as a source of functional ingredients of prebiotic and antioxidant value. *Antioxidants*, *8*(9), 406.

González-Félix, M. L., Gatlin III, D. M., Urquidez-Bejarano, P., de la Reé-Rodríguez, C., Duarte-Rodríguez, L., Sánchez, F., Casas-Reyes, A., Yamamoto, F. Y., Ochoa-Leyva, A., & Perez-Velazquez, M. (2018). Effects of commercial dietary prebiotic and probiotic supplements on growth, innate immune responses, and intestinal microbiota and histology of *Totoaba macdonaldi*. *Aquaculture*, *491*, 239–251.

Guerreiro, I., Serra, C. R., Pousão-Ferreira, P., Oliva-Teles, A., & Enes, P. (2018). Prebiotics effect on growth performance, hepatic intermediary metabolism, gut microbiota and digestive enzymes of white sea bream (*Diplodus sargus*). *Aquaculture Nutrition*, *24*(1), 153–163.

Hashemzadeh-Cigari, F., Khorvash, M., Ghorbani, G. R., Kadivar, M., Riasi, A., & Zebeli, Q. (2014). Effects of supplementation with a phytobiotics-rich herbal mixture on performance, udder health, and metabolic status of Holstein cows with various levels of milk somatic cell counts. *Journal of Dairy Science*, *97*(12), 7487–7497.

Hassaan, M. S., Soltan, M. A., Jarmołowicz, S., & Abdo, H. S. (2018). Combined effects of dietary malic acid and *Bacillus subtilis* on growth, gut microbiota and blood parameters of Nile tilapia (*Oreochromis niloticus*). *Aquaculture Nutrition*, *24*(1), 83–93.

Hassaan, M. S., Wafa, M. A., Soltan, M. A., Goda, A. S., & Mogheth, N. M. A. (2014). Effect of dietary organic salts on growth, nutrient digestibility, mineral absorption and some biochemical indices of Nile tilapia, *Oreochromis niloticus* L. Fingerlings. *World Applied Sciences Journal*, *29*(1), 47–55.

Hoseinifar, S. H., Ringø, E., Shenavar Masouleh, A., & Esteban, M. Á. (2016). Probiotic, prebiotic and synbiotic supplements in sturgeon aquaculture: A review. *Reviews in Aquaculture*, *8*(1), 89–102.

Jin, Y., Liu, F., Tian, L., Liu, Y., Li, S., & Cheng, J. (2016). Effect of dietary alanine and glycine supplementation on growth performance, body composition and apparent nutrient digestibility of juvenile grass carp (*Ctenopharyngodon idella*). *Israeli Journal of Aquaculture-Bamidgeh*, *68*.

Kalač, P. (2009). Chemical composition and nutritional value of European species of wild growing mushrooms: A review. *Food Chemistry*, *113*(1), 9–16.

Khatoon, N., Sengupta, P., Homechaudhuri, S., & Pal, R. (2010, December). Evaluation of algae based feed in goldfish (*Carassius auratus*) nutrition. In *Proceedings of the Zoological Society* (Vol. 63, pp. 109-114). Springer-Verlag.

Kumar, N. P. B., Mahaboobi, S., & Akhilesh, T. (2016). Effect of feed additives on growth performance of fish. *Journal of FisheriesSciences.com*, *10*(3), 84.

Kyaw, K., Koshio, S., Ishikawa, M., Yokoyama, S., Kikuchi, K., Laining, A., Muraoka, Y., & Akimoto, Y. (2010). Effects of different levels of dietary mussel phospholipids on the performances of Juvenile Kuruma Shrimp *Marsupenaeus japonicus* Bate. *Aquaculture Science*, *58*(1), 75–83.

Lee, B. K., & Kim, J. K. (2001). Production of *Candida utilis* biomass on molasses in different culture types. *Aquacultural Engineering*, *25*(2), 111–124.

Li, P., Mai, K., Trushenski, J., & Wu, G. (2009). New developments in fish amino acid nutrition: Towards functional and environmentally oriented aquafeeds. *Amino Acids*, *37*, 43–53.

Liao, I. C. (1977). On completing a generation cycle of the grey mullet, *Mugil cephalus*, in captivity. *Journal of the Fisheries Society of Taiwan*, *5*(2), 1–10.

Lim, C., Luckstadt, C., Webster, C. D., & Kesius, P. (2015). Organic acids and нeir salts. In: *Dietary Nutrients, Additives and Fish Health*, Wiley Blackwell: Canada, pp. 305–320.

Lin, S., Mai, K., & Tan, B. (2007). Effects of exogenous enzyme supplementation in diets on growth and feed utilization in tilapia, *Oreochromis niloticus* x *O. aureus*. *Aquaculture Research*, *38*(15), 1645–1653.

Lin, H., Tan, X., Zhou, C., Niu, J., Xia, D., Huang, Z., Wang, J., & Wang, Y. (2015). Effect of dietary arginine levels on the growth performance, feed utilization, non-specific immune response and disease resistance of juvenile golden pompano *Trachinotus ovatus*. *Aquaculture*, *437*, 382–389.

Lindequist, U., Niedermeyer, T. H., & Jülich, W. D. (2005). The pharmacological potential of mushrooms. *Evidence-based Complementary and Alternative Medicine*, *2*, 285–299.

Ma, M., & Hu, Q. (2023). Microalgae as feed sources and feed additives for sustainable aquaculture: Prospects and challenges. *Reviews in Aquaculture*, *16*(2), 818–835.

Makkar, H. P., Tran, G., Heuzé, V., Giger-Reverdi n, S., Lessire, M., Lebas, F., & Ankers, P. (2016). Seaweeds for livestock diets: A review. *Animal Feed Science and Technology*, *212*, 1–17.

Manning, B. B. (2015). Mycotoxin contamination of fish feeds. *Dietary Nutrients, Additives, and Fish Health*, 237–248.

Meena, D. K., Das, P., Kumar, S., Mandal, S. C., Prusty, A. K., Singh, S. K., Akhtar, M. S., Behera, B. K., Kumar, K., Pal, A. K., & Mukherjee, S. C. (2013). Beta-glucan: An ideal immunostimulant in aquaculture (a review). *Fish Physiology and Biochemistry*, *39*, 431–457.

Mišurcová, L. (2011). Chemical composition of seaweeds. In: *Handbook of Marine Macroalgae: Biotechnology and Applied Phycology*, pp. 171–192.

Mohammadi Gheisar, M., & Kim, I. H. (2018). Phytobiotics in poultry and swine nutrition–a review. *Italian Journal of Animal Science*, *17*(1), 92–99.

Mohan, K., Karthick Rajan, D., Muralisankar, T., Ramu Ganesan, A., Marimuthu, K., & Sathishkumar, P. (2022). The potential role of medicinal mushrooms as prebiotics in aquaculture: A review. *Reviews in Aquaculture*, *14*(3), 1300–1332.

Morais, T., Inácio, A., Coutinho, T., Ministro, M., Cotas, J., Pereira, L., & Bahcevandziev, K. (2020). Seaweed potential in the animal feed: A review. *Journal of Marine Science and Engineering*, *8*(8), 559.

Mustafa, S. A., & Al-Faragi, J. K. (2021). Supplementation of feed additives on aquaculture feeds: A review. *International Journal of Pharmaceutical Research, 13*(1).

Nates, S. F. (2016). Feed additives. *Aquafeed Formulation. (Nates SFM, Editor).* Academic Press: USA.

Navarrete, P., & Tovar-Ramírez, D. (2014). Use of yeasts as probiotics in fish aquaculture. *Sustainable Aquaculture Techniques*, *1*, 135–172.

Ng, W. K., Koh, C. B., Sudesh, K., & Siti-Zahrah, A. (2009). Effects of dietary organic acids on growth, nutrient digestibility and gut microflora of red hybrid tilapia, *Oreochromis* sp., and subsequent survival during a challenge test with *Streptococcus agalactiae. Aquaculture Research*, *40*(13), 1490–1500.

NRC (National Research Council). (2011). *Nutrient Requirements of Fish.* National Academy Press: Washington, DC, p. 376.

Pandey, A. K., Kumar, P., & Saxena, M. J. (2019). Feed additives in animal health. *Nutraceuticals in Veterinary Medicine*, pp. 345–362.

Panigrahi, A., & Azad, I. S. (2007). Microbial intervention for better fish health in aquaculture: The Indian scenario. *Fish Physiology and Biochemistry*, *33*, 429–440.

Pascual, M. M., Hualde, J. P., Bianchi, V. A., Moreno, P., Castro, J. M., & Luquet, C. M. (2018). Wheat grains fermented by fungal mycelia (*Pleurotus ostreatus* or *Lentinus edodes*) as alternative feed ingredients for juvenile rainbow trout (*Oncorhynchus mykiss*). *Aquaculture International*, *26*(6), 1339–1352.

Rachmawati, D., Samidjan, I., Elfitasari, T., Nurhayati, D., & Yuniarti, T. (2023). Dietary methionine improves feed efficiency, growth performance, body composition and digestive enzyme activity of catfish (*Pangasianodon hypophthalmus*) fingerlings. *AACL Bioflux*, *16*(3), 1805–1815.

Rajauria, G. (2015). Seaweeds: A sustainable feed source for livestock and aquaculture. In: *Seaweed Sustainability*. Academic Press, pp. 389–420.

Ren, G., Xu, L., Lu, T., Zhang, Y., Wang, Y., & Yin, J. (2019). Protective effects of lentinan on lipopolysaccharide induced inflammatory response in intestine of juvenile taimen (*Hucho taimen*, Pallas). *International Journal of Biological Macromolecules*, *121*, 317–325.

Ringø, E., & Song, S. K. (2016). Application of dietary supplements (synbiotics and probiotics in combination with plant products and β-glucans) in aquaculture. *Aquaculture Nutrition*, *22*(1), 4–24.

Ringø, E., Li, X., Doan, H. V., & Ghosh, K. (2022). Interesting probiotic bacteria other than the more widely used lactic acid bacteria and bacilli in finfish. *Frontiers in Marine Science*, *9*, 848037.

Ringø, E., Løvmo, L., Kristiansen, M., Bakken, Y., Salinas, I., Myklebust, R., Olsen, R. E., & Mayhew, T. M. (2010). Lactic acid bacteria vs. pathogens in the gastrointestinal tract of fish: A review. *Aquaculture Research*, *41*(4), 451–467.

Safavi, S. V., Kenari, A. A., Tabarsa, M., & Esmaeili, M. (2019). Effect of sulfated polysaccharides extracted from marine macroalgae (*Ulva intestinalis* and *Gracilariopsis persica*) on growth performance, fatty acid profile, and immune response of rainbow trout (*Oncorhynchus mykiss*). *Journal of Applied Phycology, 31*, 4021–4035.

Sartori, S. B., Ferreira, L. F. R., Messias, T. G., Souza, G. D., Pompeu, G. B., & Monteiro, R. T. R. (2015). Pleurotus biomass production on vinasse and its potential use for aquaculture feed. *Mycology*, *6*(1), 28–34.

Sayed, S. F., & Ahmed, I. (2022). Effects of dietary phenylalanine: Tyrosine ratio on growth, DNA/RNA, serum biochemistry, digestive enzyme activities and physiological responses of *Heteropneustes fossilis*. *Aquaculture Studies*, *23*(1), AQUAST859.

Schulz, E., & Oslage, H. J. (1976). Composition and nutritive value of single-cell protein (SCP). *Animal Feed Science and Technology*, *1*(1), 9–24.

Sharf, Y., & Khan, M. A. (2020). Effect of dietary isoleucine level on growth, protein retention efficiency, haematological parameter, lysozyme activity and serum antioxidant status of fingerling *Channa punctatus* (Bloch). *Aquaculture Nutrition*, *26*(3), 908–920.

Siddik, M. A., Sørensen, M., Islam, S. M., Saha, N., Rahman, M. A., & Francis, D. S. (2023). Expanded utilisation of microalgae in global aquafeeds. *Reviews in Aquaculture*, *16*(1), 6–33.

Song, S. K., Beck, B. R., Kim, D., Park, J., Kim, J., Kim, H. D., & Ringø, E. (2014). Prebiotics as immunostimulants in aquaculture: A review. *Fish & Shellfish Immunology*, *40*(1), 40–48.

Spolaore, P., Joannis-Cassan, C., Duran, E., & Isambert, A. (2006). Commercial applications of microalgae. *Journal of Bioscience and Bioengineering*, *101*(2), 87–96.

Tan, X., Lin, H., Huang, Z., Zhou, C., Wang, A., Qi, C., & Zhao, S. (2016). Effects of dietary leucine on growth performance, feed utilization, non-specific immune responses and gut morphology of juvenile golden pompano *Trachinotus ovatus*. *Aquaculture*, *465*, 100–107.

Tang, L., Feng, L., Sun, C. Y., Chen, G. F., Jiang, W. D., Hu, K., Liu, Y., Jiang, J., Li, S. H., Kuang, S. Y., & Zhou, X. Q. (2013). Effect of tryptophan on growth, intestinal enzyme activities and TOR gene expression in juvenile Jian carp (*Cyprinus carpio* var. Jian): Studies in vivo and in vitro. *Aquaculture*, *412*, 23–33.

Thatoi, H., & Singdevsachan, S. K. (2014). Diversity, nutritional composition and medicinal potential of Indian mushrooms: A review. *African Journal of Biotechnology*, *13*(4), 523–545.

Tian, Y., Guo, C., Zhang, X., Xie, S., Zhou, Q., Luo, J., Zhu, T., Yang, Y., Li, X., & Jin, M. (2023). Effects of dietary histidine level on growth, antioxidant capacity and TOR signaling pathway in juvenile swimming crabs, *Portunus trituberculatus*. *Aquaculture Reports*, *33*, 101869.

Van Doan, H., Doolgindachbaporn, S., & Suksri, A. (2016). Effects of Eryngii mushroom (*Pleurotus eryngii*) and *Lactobacillus plantarum* on growth performance, immunity and disease resistance of Pangasius catfish (*Pangasius bocourti*, Sauvage 1880). *Fish Physiology and Biochemistry*, *42*, 1427–1440.

Van Doan, H., Hoseinifar, S. H., Esteban, M. Á., Dadar, M., & Thu, T. T. N. (2019). Mushrooms, seaweed, and their derivatives as functional feed additives for aquaculture: An updated view. *Studies in Natural Products Chemistry*, *62*, 41–90.

Van Doan, H., Soltani, M., & Ringø, E. (2021). In vitro antagonistic effect and in vivo protective efficacy of Gram-positive probiotics versus Gram-negative bacterial pathogens in finfish and shellfish. *Aquaculture*, *540*, 736581.

Van Hai, N. (2015). Research findings from the use of probiotics in tilapia aquaculture: A review. *Fish & Shellfish Immunology*, *45*(2), 592–597.

Wan, A. H., Davies, S. J., Soler-Vila, A., Fitzgerald, R., & Johnson, M. P. (2019). Macroalgae as a sustainable aquafeed ingredient. *Reviews in Aquaculture*, *11*(3), 458–492.

Wang, L. M., Zhu, W. B., Yang, J., Miao, L. H., Dong, J. J., Song, F. B., & Dong, Z. J. (2018). Effects of dietary cystine and tyrosine on melanogenesis pathways involved in skin color differentiation of Malaysian red tilapia. *Aquaculture*, *490*, 149–155.

Xie, S. W., Tian, L. X., Li, Y. M., Zhou, W., Zeng, S. L., Yang, H. J., & Liu, Y. J. (2015). Effect of proline supplementation on anti-oxidative capacity, immune response and stress tolerance of juvenile Pacific white shrimp, *Litopenaeus vannamei*. *Aquaculture*, *448*, 105–111.

Yarnold, J., Karan, H., Oey, M., & Hankamer, B. (2019). Microalgal aquafeeds as part of a circular bioeconomy. *Trends in Plant Science*, *24*(10), 959–970.

Yeh, T. P. (1963). Nutrient composition tables of feed stuffs in Taiwan. *Journal of the Chinese Agricultural Chemical Society*, *1*, 56.

Yilmaz, S., Yilmaz, E., Dawood, M. A., Ringø, E., Ahmadifar, E., & Abdel-Latif, H. M. (2022). Probiotics, prebiotics, and synbiotics used to control vibriosis in fish: A review. *Aquaculture*, *547*, 737514.

Zhang, T., Chi, Z., & Sheng, J. (2009). A highly thermosensitive and permeable mutant of the marine yeast *Cryptococcus aureus* G7a potentially useful for single-cell protein production and its nutritive components. *Marine Biotechnology*, *11*, 280–286.

Zhang, J., Hu, Y., Ai, Q., Mao, P., Tian, Q., Zhong, L., Xiao, T., & Chu, W. (2018). Effect of dietary taurine supplementation on growth performance, digestive enzyme activities and antioxidant status of juvenile black carp (*Mylopharyngodon piceus*) fed with low fish meal diet. *Aquaculture Research*, *49*(9), 3187–3195.

Zhang, S., Wang, C. A., Liu, S., Wang, Y., Lu, S., Han, S., Jiang, H., Liu, H., & Yang, Y. (2023). Effect of dietary phenylalanine on growth performance and intestinal health of triploid rainbow trout (*Oncorhynchus mykiss*) in low fishmeal diets. *Frontiers in Nutrition*, *10*, 1008822.

Zhao, X., Yang, Y., Wang, L., Chen, H., Zhu, L., Li, L., & Kong, X. (2022). Aspartate induces metabolic changes and improves the host's ability to fight against *Aeromonas hydrophila* infection in *Cyprinus carpio*. *Aquaculture Research*, *53*(3), 1050–1061.

Zou, H. K., Hoseinifar, S. H., Miandare, H. K., & Hajimoradloo, A. (2016). *Agaricus bisporus* powder improved cutaneous mucosal and serum immune parameters and up-regulated intestinal cytokines gene expression in common carp (*Cyprinus carpio*) fingerlings. *Fish & Shellfish Immunology*, *58*, 380–386.

CHAPTER 7

The Concept of Phytotherapy in Aquaculture Systems

RAJINDER KAUR[1*], P. S. TANWAR[1], and TARANG KUMAR SHAH[2]

[1]*Krishi Vigyan Kendra, Barnala, Guru Angad Dev Veterinary and Animal Sciences University, Ludhiana, Punjab, India*

[2]*Department of Fisheries, DGCN College of Veterinary and Animal Sciences, CSKHPKV, Palampur, Himachal Pradesh, India*

*Corresponding author

ABSTRACT

Fish production needs to be increased primarily in the aquaculture sector as it provides a proteinaceous, safe, reliable, and economical source of food to the large population of the world. Cultural practices have shifted from extensive to intensive, resulting in increased fish production. However, the increase in production is restricted by various factors and challenges, among which the increasing number of infectious diseases is having the major impact. Antibiotics and disease-controlling chemicals are some of the measures that can possibly mitigate diseases in aquaculture systems. These practices pose a great threat to human and environmental health. Over the past several years, various advanced curative approaches have been used to reduce the devastating/harmful effects caused by chemicals and drugs. Among these approaches, herbal medicines are used as an alternative as plants (in the case of herbs) and their extracts have shown promising results as antimicrobial agents, growth promoters, and immunostimulants. However, this practice is still emerging and is currently confined only to laboratory animals. Various executed studies have concluded the efficacy

Advances in Freshwater Aquaculture: Technology for Sustainable Development. Tarang Kumar Shah, Madhu Sharma, Mohd. Ashraf Rather, & Avdhesh Kumar (Eds.)

DOI: 10.1201/9781779640857-7

of herbal plants and their extracts against the infectious diseases in the aquaculture sector.

7.1 INTRODUCTION

As per the reports of the Food and Agriculture Organization (FAO), the aquaculture sector is growing rapidly among all the other food-producing sectors, presently contributing to about 50% of the world's total fish used for nutrition. Human population is accelerating at a higher rate, which imposes huge pressure on aquaculture resources to provide safe, reliable, proteinaceous, and economic supply of food. In efforts of fulfilling the per capita fish consumption demand, world aquaculture has shifted from extensive to intensive systems which are based on high stocking densities and depend on artificially formulated feed, chemicals, probiotics, and drugs. The success rate of aquaculture systems mainly depends on the higher survival rate of stocked fish and its overall growth.

The major constraint in the growth of the aquaculture sector is the occurrence of fish diseases. In the case of severity of pathogen attacks, culturists may face serious financial losses. The outbreak of contagious diseases is generally stimulated by ecological changes that are basically due to human interventions such as advancement in agricultural practices or technology, industrialization, and environmental degradation (Rahimi et al., 2022). The indiscriminate use of antibiotics may increase various side effects like antimicrobial resistance and the residues may lead to serious health issues to consumers as well.

Aquaculturists have to rely on antibiotics, synthetic drugs, chemotherapeutants, and pesticides to fight against the emergence of diseases in the aquaculture sector. In aquaculture, antibiotics and chemical drugs are either administered through feeds or added directly in the water. Studies disclosed that a major part (70–80%) of the antibiotics given to fish enter the water through excretion (Rigos et al., 2004; Cabello et al., 2013; Burridge et al., 2010). These drugs coupled with other environmental factors create a selective pressure on the aquatic microbes and trigger them to undergo selective mutations to survive under unfavorable conditions. Moreover, fish feed is a known source of antimicrobial resistant bacteria (Economou and Gousia, 2015).

In aquaculture, feed wastage is comparatively more than in other livestock husbandry practices, which increases more chances of horizontal gene transfer (HGT) through natural transformation, transduction, or conjugation in the environment, thus resulting in the formation of antibiotic resistant bacteria. Such systems are symbolized as "genetic hotspots" for gene transfer. Thus, with increasing dependence of human population on the aquaculture sector, it is necessary that we should identify the suitable alternatives for antibiotics that could be applied through feed or water which may reduce the chances of developing the antimicrobial resistance and stimulate the immune system of fish.

The emergence of antibiotic resistance has raised the need to combat the issue and develop sustainable aquaculture practices that are based on using approaches such as probiotics, essential oils, and natural plant sources that enhance the innate immunity and reduce the chances of disease emergence (Sihag and Sharma, 2012; Romero Ormazabal et al., 2012, Harikrishnan et al., 2011a). Nowadays vaccination seems to be cumbersome and also a difficult alternative to treat diseases.

7.2 DISEASE OUTBREAKS THREATEN THE AQUACULTURE INDUSTRY

Disease outbreaks are among the major concerns that impede the productivity of culture systems and substantially result in economic loss to the aquaculture industry. The intensification of culture systems has also resulted in the increase loss of fish production in pond ecosystems in cases of disease occurrence with the loss accounting for approximately 60% (Bondad-Reantaso et al., 2005). The FAO stated that globally, the occurrence of infectious diseases in the aquaculture industry has led to a huge loss of over US$6 billion per annum (Brummett et al., 2014). In recent years, various cases of disease outbreaks have originated. In 2022, virus outbreak worried shrimp farmers in Ernakulam district, Kochi, India. The outbreak was of white spot syndrome virus (WSSV) that affected about 4000 ha of aquaculture area in the district.

In 2023, tilapia producers in Colombia lost around 5% of their 120,000 metric tons of annual production following an outbreak of *Streptococcus agalactiae* ST-7 Ia bacterium in a large number of farming areas. In the same year, scale drop disease virus, which is a deadly fish virus, attacked sea bass farming in Singapore. This put a halt to its commercial production. The outbreak occurred amid efforts to enhance the native seafood production with an objective to fulfill 30% of the nutritional need of the country's population by 2030.

For maintaining a pathogen/disease-free environment in aquafarms, the indiscriminate use of chemotherapeutants, antibiotics, and pesticides with a lack of best management practices has resulted in a negative impact in aquaculture production industry (Cabello, 2006; Cabello et al., 2013; Economou and Gousia, 2015; Chuah et al., 2016). The surfacing of new disease-causing pathogens, development of drug-resistant microbes, and bioaccumulation of residues of used antibiotics and pesticides in the flesh of finfish and shellfish are some of the most serious negative impacts of using chemical therapeutics.

7.3 ENTRY ROUTE INTO THE ENVIRONMENT

Scientific reports have mentioned that approximately 75% of the antibiotics given by the feed administration method enter the environment through either the metabolic progression of "secretion and excretion" from the reared organisms or leaching from uneaten antibiotic administered feed (Lalumera et al., 2004). The increasing use of antibiotic in aquaculture and its deleterious results on the aquatic ecosystems have caught the attention of scientists as its presence is very much deleterious to microbes and primary producers. Santos et al. (2010), Zounkova et al. (2011), Ahmad et al. (2022), and Wollenberger et al. (2000) studied the long-term consequences of oxalinic acid exposure in daphnids and reported the disruptive reproduction in the tested organisms even at lower concentrations.

7.4 NEED OF AN ALTERNATE CURATIVE METHOD

About 90% of the aquaculture production is being practiced in developing countries, where regulations and enforcements on indiscriminate application of antibiotics and chemicals are totally absent or not strictly implemented (Chuah et al., 2016). The absence of proper regulations has led to the indiscriminate use of antibiotics and various other chemicals. In aquaculture, the routine use of antibiotics as growth promoters is not reported (Burridge et al., 2010). However, in rearing shrimp and salmon species in the aquaculture industry, the use of antibiotics is a routine for prophylaxis and growth promoter (Holmstrom et al., 2003; Buschmann et al., 2012).

Despite the occurrence of HGT of resistance genes from farmed fish species and their surroundings to human population, the aquaculture supply chain may function as an undermined route for the spread of antimicrobial-resistant bacteria (Schar et al., 2021). The bacteria can colonize a new ecosystem after passively migrating more than 50 km through the water column (Cantrell et al., 2020; Kragesteen et al., 2018). Preena et al. (2020) stated that antimicrobial resistance brought up by aquaculture practices can therefore spread to microbial strains in hydrodynamically connected ecosystems that are therapeutically relevant. According to a study on the Mekong Delta's riverine ecosystem, aquaculture systems promote the spread and survival of genes that confer resistance to antibiotics. Therefore, using naturally occurring phytocompounds derived from plants as a treatment is crucial. Plants are extensively capable of treating human illnesses, and their extracts are verified to be highly beneficial in boosting both human and animal immune systems, with significant potential for use in the fishing industry (Wang et al., 2017).

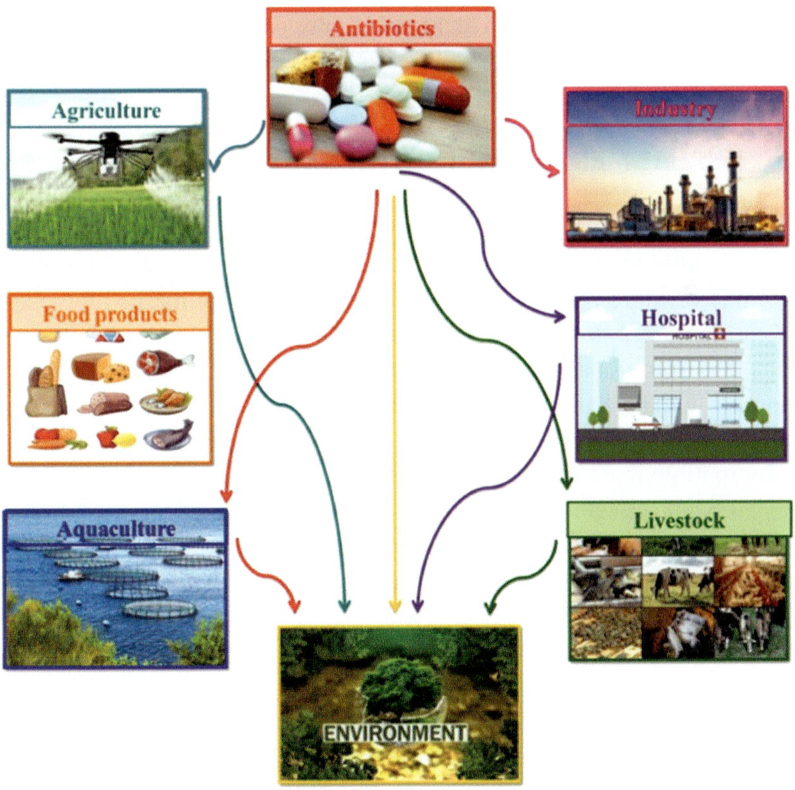

FIGURE 7.1 Entry routes of antibiotics in the environment.

Source: Cherian et al. (2023).

Besides using antibiotics and chemotherapeutants in aquaculture, there are various substitution approaches to improve the sustainability and worth of produce from aquaculture production systems. Phytotherapy has proven to be quite handy and a viable alternative method to chemotherapy. Several researchers have reported the beneficial properties of plant-based medicines as antistress, growth promoters, appetizers, tonic, and immunostimulants (Chakraborty and Hancz, 2011; Harikrishnan et al., 2011b). Herbal medicines in comparison to synthetic drugs are cost-effective, environment and farmer friendly, easily biodegradable, no or less residual impact, nonhabit forming, nonnarcotic, nonresistance forming, and without major side effects (Raman, 2017). The use of phototherapy in aquaculture is a preferable approach over using synthetic chemicals because it has low environmental risks and side effects on fish and human health and even more, the raw materials are less expensive, renewable, biodegradable, easily available, labor extensive, and less harmful to humans (Raman, 2007, 2017).

7.5 HERBAL MEDICINE/PHYTOTHERAPY

Herbal medicines are plant derived or prepared with therapeutic or other animal health advantage and contain either raw or processed ingredients from one or more plants. The end of the 18th century marked the use of herbal medicines in healthcare. According to modern medicine, Rasayan drugs are traditionally used against a plethora of seemingly diverse disorders with no pathophysiological connections. In the early 19th century, the use of botanical medicines started as they were biodegradable and biocompatible in nature. Fortunately, nothing has been reported regarding the pathogen resistance against herbal therapy. Some of the herbal plants have the property to fight against microbial attacks as they tend to possess antiviral, antibacterial, and antifungal properties. Thereafter, search has been to develop advanced and efficient alternative therapy from renewable and biodegradable resources.

Medicinal plants manufacture secondary metabolites (SMs) that are basically produced in reaction of self-defense. These released organic chemicals are categorized as antimicrobial, bacteriostatic, chemotherapeutic, and bactericidal based on their mechanism of function (Purohit and Mathur, 1999). Herbal medicines and plant extracts are referred to as "Rasayan" in homeopathic treatment which have already brought into use since time immemorial to serve and manage diverse microbial diseases and are immensely efficient in treating various diseases in fish (Raman, 2001; 2004; 2007; Raman and Rahman, 2002; Raman et al., 2004; Ali et al., 2000; Digrak et al., 1999, 2001; Koneri et al., 2006; Parvati and Kumar, 2002).

Furthermore, we will discuss in brief about the origin of phytotherapy and its use as antimicrobial agents in the fishing industry to combat various lethal fish diseases.

7.6 HISTORY OF PHYTOTHERAPY

The word "therapy" comes from the modern Latin word therapia, originally from Greek therapeia, which means to treat medically. The Latin prefix phyto stands for plant and is called photon in Greek. In other words, "phytotherapy" is basically known for herbal medicine. Thus, phytotherapy means using plant-derived medications as treatment and also for the prevention of disease.

Worldwide, phytotherapy has a long history in treating human and animal diseases since time immemorial. In countries such as China, India, as well as Southeast Asian and few South and Central America countries, phytotherapy is considered mainstream, while in Western countries, naturopathy and herbal medicine are increasingly acknowledged (Kolkovski and Kolkovski, 2011). The term "phytotherapy" was first used in 1913 by French physician Henri Leclerc. Thereafter, it was used in English language with its collective definition in 1934. In 1960, German herbalist Rudolf Fritz Weiss published a book on phytotherapy entitled *Lehrbuch der Phytotherapie* (1960, Herbal Medicine). Medicinal herbs and their extracts have been actively used as medicines and their active ingredients possibly relieved the symptoms and were even beneficial in curing the diseases (Rodrigues et al., 2012).

Various experiments have been carried out to test the use of plant-derived materials having antimicrobial properties to reduce the impact or control parasitic infections in human beings, fish, and animals (Abd El-Galil and Aboelhadid, 2012). The primary concept behind using herbal medicines in aquaculture systems is to reduce the use of chemical parasiticides and therapeutants, which ultimately provides superior sustainability for fish production, fewer environmental issues, and possess properties to fight against and treat fish diseases, thus resulting in fewer or no mortality and reduced financial loss. Second, the use of phytotherapics is less prone to the development of disease-resistant fish (Kulkarni et al., 2013) as their mode of action affects various targets simultaneously (Bakkali et al., 2008).

Currently worldwide, more than 25% of the prescribed modern medicines are made from or their basic origin is from medicinal plants. As per the data reported by the World Health Organization in 2004, it was estimated that around 80% of the world's population residing in the developing countries of Africa, Asia, and Latin America consider herbal medicines for primary health treatment. Planet Earth is a home to over 750,000 plants and among them, biological activities of only 5% of the plants' species have been tested so far. India is one of the world's 12 leading biodiversity rich countries glorified with 16 agroclimatic zones that are blessed with 45,000 different species of flora and 15,000 medicinal plants that included 7000 plants sourced by Ayurveda, 700 by Unani medicine, 600 by Sidha, 450 by homoeopathy, and 30 by modern medicine (Vaidyaratnam, 1995). Thus, many plants and herbs can be assessed for their potency against fish pathogens that can act as suitable alternatives to chemical drugs and antibiotics in aquaculture.

7.6.1 POSITIVE IMPACT OF HERBAL DRUGS

Due to increase in the emergence of antibiotic-resistant pathogens and antibiotic residues, people are acquainted with the adverse effects and consequences of using antibiotics (Chong et al., 2020). Indirect discharge through excretion is also harmful for the environment and microbial community. Hence, shifting toward a natural and efficient approach that could provide promising results is being considered in aquaculture and the use of herbal medicines is an alternative approach. These medicines are affordable and safer to use when compared with synthetic growth promoters and have promising results in aquatic organisms.

In simple words, farm-reared fish can be healthier and can grow bigger in size depending on herb-administered feeds. The application of phytotherapy in aquaculture has shown positive results in enhancing the growth and the treated fish has healthier biological responses. The study conducted by Goda (2008) reported enhanced growth, higher diet utilization efficiency, and hematological indices on feeding ginseng herb supplemented feed to *Oreochromis niloticus*.

7.7 APPLICATIONS OF PHYTOTHERAPY IN AQUACULTURE

The herbs can be used in various forms such as fresh herbs, dried, powdered, extracts (in solvents with different properties such as alcohol, water, ether, acetone), juices, or essential oils. The phytomedicines are applied in various ways as decoctions (individual) or concoctions (mixed) or in combination with other drugs for enhancing the efficiency in fish health management (Harikrishnan, 2003). The most commonly used and studied worldwide herb is Neem (*Azadirachta indica*), also popular as "sarva roga nivarak" in Ayurveda. It literally means the "healer of all ailments" and in the ancient time, people of India designated neem tree as "Village Pharmacy." While considering the medicinal values and facts of Neem, United Nations has declared it as the "Tree of the 21st Century" and its medicinal effects are deeply studied and reported in scientific research papers. The water soluble part of alcoholic extract of *A. indica* leaves possessed antifertility, hepatoprotective, hypoglycemic, hypolipidemic, hypotensive, and antiserotonin activity (Chattopadhyay, 1991; Chattopadhyay et al., 1992; Sun and Wang, 1992; Chattopadhyay, 1995; Kirtikar and Basu, 1933). Besides neem, many other herbs such as *Adhatoda vasica*, turmeric (*Curcuma longa*), garlic (*Allium sativum*), and tulsi (*Ocimum sanctum*) have been tested against aquacultural diseases (Raman, 2004). Even China (leading country in fish production) is reaping the benefits of phytotherapy as treatment against various finfish and shellfish diseases.

Phytotherapy is a branch of medical practice that mainly focuses on using traditional methods of medication rather than modern approaches as it mainly depends on medicinal herbalism. The application of phytotherapy in aquaculture is rather new but is slowly gathering the attention of scientists and aquaculturists as an alternative therapy or a curative method to the use of synthetic drugs.

Diversified bioactive compounds can be extracted from herbal plants that are basically produced by plants in response to their habitats, unfavorable or rough weather conditions, and pathogenic attack, and these are often referred to as SMs, alkaloids, essential oils, phenolics, saponins, terpene, and carbohydrates synthesized by plants. Although carbohydrates are the primary metabolites, glycosylation is the process that transforms the primary metabolites into SMs that are able to modulate protein–protein interactions such as recognition, binding, catalytic activity, and turnover. All these have given rise to an array of immunoprotective effects for the treatment of nonorigin hosts, including immunostimulants, antibacterial, antiviral, antifungal, and antistress activities (Table 7.1).

Studies have unveiled the importance of having a strong immune system in commercial fish species so that mortality can be minimized during a disease outbreak. Herbal medicines are having the potency to boost the immune system of host in response to disease occurrence for treatment. These days, phytotherapeutic agents are synthesized in combination with natural ingredients, especially with biocompounds that are naturally present in plants. Immunostimulants are given primarily to fish to stimulate its defense mechanism so that fish can become resistant to diseases.

Antibacterial agents are generally used for controlling infectious disease caused by microbes such as *Aeromonas hydrophila*, *Vibrio parahaemolyticus*, and *Vibrio damsela*. Antiviral agents are used against fish viruses such as WSSV, salmonid rhabdovirus, and viral hemorrhagic septicemia virus. Antifungal agents are used for treating diseases related to fungal pathogens such as *A. hydrophila*, *Aspergillus flavus*, and *Fusarium oxysporum*. Farmed fish are often affected by ambient abiotic and biotic stressors that can alter the metabolic state and eventually affect the health profile. The stress level of fish can be diminished by harnessing antistress agents.

7.7.1 ANTIBACTERIAL PROPERTY

Vast number of publications have reported the certainty of various herbs that could instigate the fish immune system and provide resistance against bacterial pathogens. Dietary supplementation of Nile tilapia with *A. sativum*, *Lonicera japonica*, *Echinacea purpurea*, and *Astragalus membranaceus* resulted in inducing the immune system and improving the survival rate when challenged with *A. hydrophila* (Shalaby et al., 2006; Aly et al., 2008; Aly and Mohamed, 2010; Ardo et al., 2008; Pachanawan et al., 2008). Turmeric (*C. longa*) extract boast immune system and provide protection to fish against *Aeromonas* and *Staphylococcus* (Hota and Dey, 1997).

Oil extracted from the nonedible parts of neem *A. indica* such as leaves, bark, and seeds possess a wide spectrum of antibacterial action against Gram-negative and Gram-positive microorganisms, including Streptomycin-resistant strains and *Mycobacterium tuberculosis* (Chopra et al., 1952). Direkbusarakom's study has proved that *Momordica charantia* is the most effective herb against fish and shrimp pathogenic bacteria (Direkbusarakom et al., 1998).

In the pond culture system, Das et al. (1999) recommended a dose of 10 ppm of aqua-neem to control bacterial diseases such as Hemorrhagic septicemia, bacterial gill disease, dropsy, and tail and fin rot in fin fish. Aqueous extracts of Chinese herbal medicines were also found to be efficient as antibacterial agents against shrimp bacterial pathogens (Xu et al., 1993). Castro et al. (2008) reported that 31 methanolic extracts of Brazilian plants showed antibacterial activity in agar diffusion assay on testing against fish pathogenic bacteria *A. hydrophila*, *S. agalactiae*, and *Flavobacterium columnare.* Among the three bacteria, *F. columnare* was most susceptible to the most tested methanolic extracts.

Wei and Musa (2008) found the minimum inhibitory concentration of tested concentrations of aqueous extract of garlic (500, 250, 125, 62.5 mg/mL) against four Gram-negative and two Gram-positive and 18 isolates of *Edward siella tarda.* All the tested concentrations were found to be effective against the tested bacteria. *Achyrantes aspera* and *Withania somnifera*-enriched diet was fed to *Labeo rohita* and reduction in mortality was observed by 41% and 49%, respectively, when challenged with *A. hydrophila* when compared with control groups (Sharma et al., 2010; Vasudeva et al., 2006). There are large numbers of other plants possessing antibacterial properties which can be tested and used against pathogenic bacteria in fish and shellfish.

7.7.2 ANTIVIRAL PROPERTY

Few of the studies conducted to evaluate the efficacy of herbal plants against viral pathogens of shrimp and fish are mentioned below. The two commonly found herbs *Cynodon dactylon* and *M. charantia* exhibit antiviral activity against WSSV and human vaccinia virus. Several species of herbs and plants showed a strong antiviral action against WSSV on extracting with organic solvents such as ether, chloroform, and ethanol. Citarasu et al. (2006) assessed the effect of methanolic extracts of Indian

TABLE 7.1 Antimicrobial Activities Present in Different Plant Extracts

S. No	Scientific Name	Common Name	Part Used	Properties	Reference
1.	*Aloe barbadensis*	*Aloe vera*	Leaf	Antimicrobial (including antifungal, antibacterial, antiviral, and anthelmintic)	
2.	*Melocanna baccifera*	Muli Bamboo	Leaf	antifungal, antibacterial, antiprotozoal, antitussive, and immunomodulatory	Khan et al. (1987)
3.	*Allium sativum*	Garlic	Tuber	Hypolipidemic, antimicrobial, antihypertensive, and hepatoprotective	
4.	*Excoecaria agallocha*	Mangrove	Leaf	Antimicrobial and immunomodulatory	
5.	*Zingiber officinale*	Zinger	Root	Antioxidants, anti-inflammatory, antimicrobial	
6.	*Mentha piperita*	Peppermint	Leaf	Antimicrobial and immunostimulant	
7.	*Arum maculatum*	Cuckoo pint	Leaves	Antimicrobial activity, antioxidant properties, antimutagenic, anticarcinogenic, and cardioprotective activities	
8.	*Thymus vulgaris*	Common thyme	Oil	Antimicrobial activity, antibacterial, antifungal, and antioxidant	
9.	*Achillea cucullata*	Gandrain	Oil	Antioxidant, antimicrobial, and enzyme-inhibition activity	
10.	*Anisomeles malabarica*	Malabar catmint	Leaves	Antioxidant and antibacterial activities	
11.	*Cynara cardunculus*	Cardoon	Leaf	Antimicrobial activity	
12.	*Thymus linearis*	Himalayan thyme	Oil	Antimicrobial, antioxidant activity, and antiseptic	
13.	*Thymus daenensis*	Thyme	Oil	Antifungal activity	
14.	*Indigofera suffruticosa*	Indian indigo	leaves	Antimicrobial activity	

medicinal plants, namely *Cyanodon dactylon*, *Aegle marmelos*, *Tinospora cordifolia*, *Picrorhiza kurooa*, and *Eclipta alba* on WSSV challenged *Penaeus monodon*. The report showed 74% survival in *P. monodon* with reduction in WSSV load in various tissues.

Krishnan et al. (2010) found the antiviral activity of dasyscyphin C (C28H40O8) extracted from leaves of *Eclipta prostrata* against fish viral nervous necrosis. Reduction in mortality was reported in olive flounder infected with the lymphocystis disease virus after intraperitoneal injection with *Punica granatum* extracts (Balasubramanian et al., 2008). Hudson et al. (2000) found a potent antiviral activity in *M. charantiais* among several screened West African plants. Further, a strong activity of *M. charantiais* against human HSV-1 (Herpes Simplex Virus) was reported by Beloin et al. (2005).

7.7.3 ANTIFUNGAL PROPERTY

Using synthetic fungicides in aquaculture systems to control fish diseases has various drawbacks due to their carcinogenicity, teratogenicity, high and acute toxicity, long degradation period, and their deleterious impact on edible items causing various health issues to human beings (Raman, 2017). In recent years, extracts of various herbal plants have been used for curing fungal diseases in fish and it is basically due to the property of its nontoxic behavior, with less or no side effects and easy availability at a cheap price.

Neem leaf extract, oil, and seed kernel were tested against fungi including *Microsporum*, *Epidermophyton*, *Trichosporon*, *Candida*, and *Geotircum* and were found to be highly efficient in treating the fungal infection (Alice and Sivaprakasam, 1999). Other herbal extracts of garlic, clove extract, turmeric, and lime (Dey and Chandra, 1994) were employed for fungal treatments in fish. A paste of garlic (2 kg), salt (2 kg), $CuSO_4$ (20 g), and $KMnO_4$ (20 g) was made and then mixed in 30–50 L of water which was further sprayed over pond surface to cure epizootic ulcerative syndrome in fish (Goswami et al., 2006). The extract of *Fructus prunusis* was tested against black gill disease in farmed shrimp (Xu et al., 1993). Harikrishnan and Balasundaram (2005) reported that *A. indica* extract is having curative effect against the fungal attack of *Aphanomyces invadans* in common carp.

Xu et al. (1993) conducted a study on Chinese herb, *F. prunusis* which was found to inhibit the growth of *F. oxysporum* which caused black gill disease in *Penaeus orientalis*. *Helenium quadridentatum* extracts were found quite efficient in reducing the impact caused by saprolegnia in fish (Auro-de-Ocampo and Jimenez, 1993).

7.7.4 ANTIHELMINTHIC AND ANTIPROTOZOAL PROPERTY

Protozoan parasites are very much common and present almost everywhere whether it is inland fisheries or marine and the fin fish and shell fish are more prone to their attack during the early stages of life cycle from egg to fry and fingerling. They are also vulnerable to the adult stage but the mortality rate is lesser when compared with the infant stage (Raman, 2017). The worms belonging to class monogenea are generally ectoparasite with no intermediate host and have significant parasitic effect on fish health (Deveney et al., 2001; Woo et al., 2002). Several organic extracts of plants such as bupleurum root, cinnamon, Chinese spice bush, and golden larch are reported to provide 100% protection to *Carassius auratus* infected with monogenean *Dactylogyrus intermedius* (Ji et al., 2012; Wu et al., 2011).

Mitchell and Hoffman (1980) reported that *Kamala* is an ancient anti-helminthic herbal medicine having the potential to remove adult cestodes and other worms from fish. According to Ayurveda, the whole neem tree is beneficial in treating helminth attack in animals as flower, oil, pulp, fruit, seed, and twig of neem are very much efficient against helminths. The ethanol extract of neem seed kernels called as Neemol was found efficient against cestode parasites attack in white leghorns (Neeraja and Narayana, 1998).

Oketch-Rabah et al. (1997) observed that the extract nyasol from *Asparagus africanus* against *Leishmania major* and *Plasmodium falciparum* can inhibit the parasites growth, whereas the Piperine derived from *Piper longum* has an antiparasitic effect on goldfish (*C. auratus*) infested with fish lice (Argulus) (Kumar et al., 2012).

Dey and Chandra (1995) reported the successful control of Trichodinosis in the fry of *Catla catla* as there was severe attack of *Trichodina indica* on catla fry infecting up to 70% of the population. The infestation came under control with the application of 1 ppm aqueous extract of garlic in 10 ppm common salt/ha/m water.

7.7.5 AS BIOPESTICIDE

Studies have reported that various plant species possess pesticidal and insecticidal properties and are reported to be highly efficient against metazoan ectoparasites. Rajasekaran et al.'s study reported that oil extracted from Tulsi (*O. sanctum*) has the potential as antimalarial drug as it possesses approximately 100% larvicidal property against the Culex mosquitoes as well as mosquito-repelling properties.

In China, Shaoqi (1989) reported an approach for successfully controlling Lernaeasis in carps by applying 400 kg/mu of fermented cow dung or pig waste or 100–150 kg/mu of distiller's dregs for every meter of water depth. Applying pine (*Pinus massoniana*) @ 20 kg per 1/15 ha water area is also a successful that can control Lernaeasis (Rath, 2000). Recently, Tiwari et al. (2006) reported that by putting stems of banana plant squarely in the fish pond with small amount of kerosene oil can biologically control the aquatic insects in nursery ponds.

7.7.6 HERBAL IMMUNOSTIMULANT

Many plant-derived compounds have found to boost the nonspecific immunostimulatory response in animals which is more evaluated in shellfish and finfish. So far, a large number of immunostimulants belonging to heterogenous group of substances such as peptides, synthetic chemicals, peptidoglycan, biological substances, glucans, nutritional factors, mycelial fungi, killed mycobacterium, cytokines, polysaccharides, muramyl dipeptide, hormones, and plant ingredients/extracts such as *Aloe vera* extract, Azadirachtin, Curcumin, Piperine, and *O. sanctum* extract have been tested and have provided a positive response to the immune system of various finfish and shellfish species (Anderson, 1992; Saurav et al., 2013; Das et al., 2015). However, the botanical products are a rich source of active substances for immunotherapy (Hadden, 1993). Herbal immunostimulants such as *Bemisia tabaci* (Srinivasan et al., 2001), Neem Azal-T/S (El-shazly and El-sharnoubi, 2000), 1-cysteine, nemicidine, nemol and vimicidine (derived from neem plant), Curcumin (derived from turmeric) (repellent Gangopadhyay, 1995), *Croton tiglium* (Babu, 1965), and Mahua oil cake (Bhatia, 1970) act as immunstimualants.

7.7.7 EFFECTS ON FISH GROWTH AND SURVIVAL

Artemia nauplii enriched with herbal products such as stressol-I and stressol-II increased the growth and efficiency significantly and decreased the osmotic stress in *Penaeus indicus* postarvae (PL 10–20) (Chitra, 1995). Herbal plants including *Eclipta erecta*, *Zingiber officinalis*, *Picrorhiza kurooa*, *Hygrophila spinosa*, *A. paniculata*, *Ocimum sacnctum*, *Phyllanthus niruri*, *Psoralea corylifolia*, *Solanum trilobatum*, *Tinospora cordifolia*, *W. somnifera*, purified Shilajit, and cod-liver oil have shown a positive impact on the growth of tested animals increasing growth, trigger the immune system and act as antistress and antibacterial in fish. Fingerlings of Nile tilapia, *O. niloticus* were fed with ginseng herb (Ginsana G115) and elevation in growth and diet utilization efficiency was observed. Furthermore, hematological indices were also tested and found to have no negative effect on utilizing ginseng herb in fish diet (Ashraf and Goda, 2008).

Extracts of herbal thyme (*Thymus vulgaris*), rosemary (*Rosmarinus officinalis*), and fenugreek (*Trigonella foenum graecum*) as feed additives to enhance growth performance were reported in European sea bass (*Dicentrarchus labrax*) (Yılmaz et al., 2006). Banana peel flour (BPF) (*Musa acuminata*) was administered in the feed of *L. rohita* at different incorporation levels ranging from 0% to 7% and observed an increase in the final weight gain and specific growth rate. The rate was higher in the fish fed with 5% BPF (Giri et al., 2016).

The effect of dietary administration of *Achyranthes aspera* extract as a natural antibiotic on growth performance of *L. rohita* reported that the herbal extract acted as immunostimulant and enhanced the growth rate as well (Jaman et al., 2017). The effect of phytogenic feed additive on the growth and susceptibility of *O. niloticus* to *A. hydrophila* infection was studied. The report revealed that optimum growth performance, feed utilization efficiency, and resistance against the tested bacteria were obtained at 0.5 g phytogenic feed additive (Silaacid® Encapsulated) kg^{-1} diet.

7.8 CONCLUSION

Synthetic antibiotics and artificial drugs are very much efficient and quick in action in controlling diseases. However, in spite of the positive role, their use results in adverse impacts on the aquatic environment and human health due to residues and are very much discouraged in the aquaculture industry. Several researchers have tested the efficacy of medicinal herbs and its extracts against fish and shell fish diseases and have found a positive impact. Moreover, they act as growth promoters, stress resistance boosters, immuno-stimulants, and prophylactic agents. This approach is very cost efficient and has less or no toxic impact on environment. Phytochemicals that are found in medicinal plants (tannins, alkaloids, and flavonoids) have shown promising antimicrobial results either by reducing the pathogen load or by eliciting the fish immunity by stimulating the innate immune system of fish through the receptor-mediated interaction. Therefore, in order to promote the use of herbal medicines, awareness programs should be conducted to popularize their use among fish farmers through result demonstration, which could help in producing healthy animal protein food for human consumption.

KEYWORDS

- **aquaculture**
- **chemicals**
- **drugs**
- **immune-stimulants**
- **organic extracts**
- **phytotherapy**

REFERENCES

Abd El-Galil, M. A., & Aboelhadid, S. M. (2012). Trials for the control of Trichodinosis and Gyrodactylosis in hatchery reared *Oreochromis niloticus* fries by using garlic. *Veterinary Parasitology*, *185*, 57–63.

Ahmad, A., Kurniawan, S. B., Sheikh Abdullah, S. R., Ahmad, R. O., & Hasan, H. A. (2022). Contaminants of emerging concern (CECs) in aquaculture effluent: Insight into breeding and rearing activities, alarming impacts, regulations, performance of wastewater treatment unit and future approaches. *Chemosphere*, *290*, 133319.

Ali, A. M., Ismail, N. H., Mackeen, M. M., Yazan, L. S., Mohamed, S. M., Ho, A. S. H., & Lajis, N. H. (2000). Antiviral, cytotoxic and antimicrobial activities of anthrquinnones isolated from the root of *Morinda elliptica*. *Pharmaceutical Biology*, *38*, 298–301.

Alice, D., & Sivaprakasam, K. (1999). Fungicidal, bactericidal and nematicidal effect of garlic clove extract. *Journal of Ecobiology*, *8*, 99–103.

Aly, S. M., & Mohamed, M. F. (2010). *Echinacea purpurea* and *Allium sativum* as immunostimulants in fish culture using Nile tilapia (*Oreochromis niloticus*). *Journal of Animal Physiology and Animal Nutrition*, *94*(5), 31–39.

Aly, S. M., Mohamed, M. F., & John, G. (2008). Echinacea as Immunostimulatory agent in Nile Tilapia (*Oreochromis niloticus*) via earthen pond experiment. In *Proceedings of the 8th International Symposium on Tilapia in Aquaculture*, pp. 1033–1041.

Anderson, D. P. (1992). Immunostimulants, adjuvants and vaccine carriers in fish: Applications to aquaculture. *Annual Review of Fish Diseases*, *2*, 281–307.

Ardo, L., Yin, G., Xu, P., Varadi, L., Szigeti, G., Jeney, Z., & Jeney, G. (2008). Chinese herbs (*Astragalus membranaceous* and *Lonicera japonica*) and boron enhance the non-specific immune response on Nile tilapia (*Oreochromis niloticus*) and resistance against *Aeromonas hydrophila*. *Aquaculture*, *275*, 26–33.

Ashraf and Goda. (2008). Effect of dietary ginseng herb (Ginsana® G115) supplementation on growth, feed utilization, and hematological indices of Nile Tilapia, *Oreochromis niloticus* (L.), fingerlings. *Journal of the World Aquaculture Society*, *30*, 205–214.

Auro-de-Ocampo, A., & Jimenez, E. M. (1993). Herbal medicines in the treatment of fish diseases in Mexico. *Veterinaria Mexico*, *24*, 291–295.

Babu, N. (1965). Observations on the toxicity of the seeds of *Croton tiglium* Linnon predatory weed fishes. *Science & Culture*, *31*, 308–310.

Bakkali, F., Averbeck, S., Averbeck, D., & Idaomar, M. (2008): Biological effects of essential oils–a review. *Food and Chemical Toxicology*, *46*, 446–475.

Balasubramanian, G., Sarathi, M., Venkatesan, C., Thomas, J., & Hameed, A. S. S. (2008). Studies on the immunomodulatory effect of extract of *Cynodon dactylon* in shrimp, *Penaeus monodon*, and its efficacy to protect the shrimp from white spot syndrome virus (WSSV). *Fish and Shellfish Immunology, 25*, 820–828.

Beloin, N., Gbeassor, M., Akpagana, K., Hudson, J., de Soussa, K., Koumaglo, K., & Arnason, J. T. (2005). Ethnomedicinal uses of *Momordica charantia* (Cucurbitaceae) in Togo and relation to its phytochemistry and biological activity. *Journal of Ethnopharmacology*, *96*, 49–55.

Bhatia, H. L. (1970). Use of mahua oil cake in fishery management. *Indian Farming*, *20*, 39–40.

Bondad-Reantaso, M. G., Subasinghe, R. P., Arthur, J. R., Ogawa, K., Chinabut, S., Adlard, R. et al. (2005). Disease and health management in Asian aquaculture. *Veterinary Parasitology, 132*, 249–272.

Brummett, R. E., Alvial, A., Kibenge, F., Forster, J., Burgos, J. M., Ibarra, R., St-Hilaire, S., Chamberlain, G. C., Lightner, D. V., Khoa, L. V., & Hao, N. V. (2014). Reducing disease risk in aquaculture. *World Bank Report Number 88257-GLB*.

Burridge, L., Weis, J. S., Cabello, F., Pizarro, J., & Bostick, K. (2010). Chemical use in salmon aquaculture: A review of current practices and possible environmental effects. *Aquaculture*, *306*, 7–23.

Buschmann, A. H., Tomova, A., Lopez, A., Maldonado, M. A., Henríquez, L. A., Ivanova, L., Moy, F., Godfrey, H. P., & Cabello, F. C. (2012). Salmon aquaculture and antimicrobial resistance in the marine environment. *PLoS One*, *7*, 42724.

Cabello, F. C. (2006). Heavy use of prophylactic antibiotic in Aquaculture: A growing problem in human and animal health and for environment. *Environmental Microbiology, 8*, 1137–1144.

Cabello, F. C., Godfrey, H. P., Tomova, A., Ivanova, L., Dolz, H., Millanao, A., & Buschmann, A. H. (2013). Antimicrobial use in aquaculture re-examined: Its relevance to antimicrobial resistance and to animal and human health. *Environmental Microbiology, 15*, 1917–1942.

Cantrell, D. L., Groner, M. L., Ben-Horin, T., Grant, J., & Revie, C. W. (2020). Modeling pathogen dispersal in marine fish and shellfish. *Trends in Parasitology*, *36*, 239–249.

Castro, S. B. R., Leal, C. A. G., Freire, F. R., Carvalho, D. A., Oliveira, D. F., Figueiredo, H. C. P. (2008). Antibacterial activity of plant extracts from Brazil against fish pathogenic bacteria. *Brazilian Journal of Microbiology*, *39*, 756–760.

Chakraborty, S. B., & Hancz, C. (2011). Application of phytochemicals as immunostimulant, antipathogenic and antistress agents in finfish culture. *Reviews in Aquaculture*, *3*, 103–119.

Chattopadhyay, R. R. (1991). Studies on alcoholic extract of leaves of *Azadirachta indica* (Beng. Neem) on carbohydrate metabolism. PhD thesis, University of Kalyani, West Bengal, India.

Chattopadhyay, R. R. (1995). Hypolipidemic activity of *Azadirachta indica* leaf extract in rats. *Proceedings of Indian National Science Academy, B*, *61*, 281.

Chattopadhyay, R. R., Sarkar, S. K., Ganguli, S., Banerjee, R. N., & Basu, T. K. (1992). Hepatoprotective activity of *Azadirachta indica* leaves on paracetamol induced hepatic damage in rats. *Indian Journal of Experimental Biology*, *30*, 738.

Chitra, S. (1995). Effect of feeding supplemented stresstol bioencapsulated Artemia franciscana on growth and stress tolerance—In *Penaeus indicus* postlarvae. MPhil dissertation, MS University, Tirunelveli

Chong, C. M., Murthy, G. A. V. S., Choy, C. Y., & Lai, K. S. (2020). Phytotherapy in aquaculture: Integration of endogenous application with science. *Journal of Environmental Biology*, *41*, 1204–1214.

Chopra, I. C., Gupta, K. C., & Nazir, B. N. (1952). Preliminary study of anti-bacterial substances from *Melia azidirachta*. *Indian Journal of Medical Research*, *40*, 511–515.

Chuah, L., Effarizah, M. E., Goni, A. M., & Rusul, G. (2016). Antibiotic application and emergence of multiple antibiotic resistance (MAR) in global catfish aquaculture. *Current Environmental Health Reports*, *3*, 118–127.

Citarasu, T., Sivaram, V., Immanuel, G., Rout, N., & Murugan, V. (2006). Influence of selected Indian immunostimulant herbs against White Spot Syndrome Virus (WSSV) infection in Black Tiger Shrimp, *Penaeus monodon* with reference to haematological, biochemical and immunological changes. *Fish and Shellfish Immunology*, *21*, 372–384.

Das, B. K., Mukherjee, S. C., Sahu, B. B., & Murjani, G. (1999). Neem (*Azadirachta indica*) extract as an antibacterial agent against fish pathogenic bacteria. *Indian Journal of Experimental Biology*, *37*, 1097–1100.

Das, R., Raman, R. P., Saha, H., & Singh, R. (2015). Effect of *Ocimum sanctum* Linn. (Tulsi) extract on the immunity and survival of *Labeo rohita* (Hamilton) infected with *Aeromonas hydrophila*. *Aquaculture Research*, *46*, 1111–1121.

Deveney, M. R., Chisholm, L. A., & Whittington, I. D. (2001). First published record of the pathogenic monogenean parasite *Neobenedenia melleni* (Capsalidae) from Australia. *Diseases of Aquatic Organisms, 46*, 79–82.

Dey, R. K., & Chandra, S. (1994). New trends in fish disease management through application of herbal material. *Fishing Chimes*, *14*, 13–14.

Dey, R. K., & Chandra, S. (1995). Preliminary studies to raise disease resistant seed (Fry) through herbal treatment of spawn. *Fishing Chimes*, *14*, 23–25.

Digrak, M., Alma, H. M., Ilcim, A., & Sen, S. (1999). Antimicrobial and antifungal effects of various commercial plants extract. *Pharmaceutical Biology*, *37*, 216–220.

Digrak, M., Alma, H. M., Ilcim, A., & Sen, S. (2001). Antimicrobial and antifungal activities of Turkish medicinal plants. *Pharmaceutical Biology*, *39*, 346–350.

Direkbusarakom, S., Ezura, Y., Yoshimizu, M., & Herunsalee, A. (1998). Efficacy of Thai traditional herb extracts against fish and shrimp pathogenic bacteria. *Fish Pathology*, *33*, 437–441.

Economou, V., & Gousia, P. (2015). Agriculture and food animals as a source of antimicrobial-resistant bacteria. *Infection and Drug Resistance*, *8*, 49–61.

El-shazly, M. M., & El-sharnoubi, E. D. (2000). Toxicity of a Neem (*Azadirachta indica*) insecticide to certain aquatic organisms. *Journal of the Egyptian Society of Parasitology*, *30*, 221–231.

Gangopadhyay, S. (1995). Use of *Curcuma longa* and *Azadirachta indica* for safe storage of food grains. In *International Conference on Current Progress in Medicinal and Aromatic Plant Research, Calcutta*, 30th December 1994–1st January 1995, p. 99.

Giri, S. S., Jun, J. W., Sukumaran, V., & Park, S. C. (2016). Dietary administration of Banana (*Musa acuminata*) peel flour affects the growth, antioxidant status, cytokine responses, and disease susceptibility of Rohu, *Labeo rohita*. *Journal of Immunology Research, 2016*, 4086591.

Goswami, B., Mondal, S., & Dana, S. S. (2006). Technical knowledge in fish farming. *Indian Journal of Traditional Knowledge*, *5*, 60–63.

Hadden, J. W. (1993). Immunostimulants. *Immunology Today*, *14*, 275–282.

Harikrishnan, R. (2003). Herbal treatment for ulcerative disease induced by *Aeromonas hydrophila* in Goldfish (*Carassius auratus*). PhD thesis, Bharathidasan University, Tiruchirapalli, India.

Harikrishnan, R., & Balasundaram, C. (2005). Modern trends in *Aeromonas hydrophila* disease management with fish. *Reviews in Fisheries Science*, *13*, 281–320.

Harikrishnan, R., Balasundaram, C., & Heo, M. S. (2011a) Diet enriched with mushroom *Phellinus linteus* extract enhances the growth, innate immune response, and disease resistanceof kelp grouper, *Epinephelus bruneus* against vibriosis. *Fish and Shellfish Immunology*, *30*, 128–134.

Harikrishnan, R., Balasundaram, C., & Heo, M. S. (2011b) Impact of plant products on innate and adaptive immune system of cultured finfish and shellfish. *Aquaculture*, *317*, 1–15.

Holmstrom, K., Graslund, S., Wahlstrom, A., Poungshompoo, S., Bengtsson, B. E., & Kautsky, N. (2003). Antibiotic use in shrimp farming and implications for environmental impacts and human health. *International Journal of Food Science Technology, 38*(3), 255–266.

Hota, S. K., & Dey, R. K. (1997). Studies on the in vitro effect of turmeric on some fish pathogenic bacteria. *Fishing Chimes*, *17*, 9–10.

Hudson, J. B., Anani, K., Lee, M. K., De Souza, C., Arnason., J. T., & Gbeassor, M. (2000). Further investigations on the antiviral activities of medicinal plants of Togo. *Pharmacological Biology*, *38*, 46–50.

Jaman, M. H. U., Khondoker, S., Ferdaus, J., Ahmed, M. T., Sana, S., Hasan, M. N., & Hasan, F. (2017). Effects of *Achyranthes aspera* to the growth performance of Rohu (*Labeo rohita*). *International Journal of Fisheries and Aquatic Studies*, *5*(3), 109–113.

Ji, J., Lu, C., Kang, Y., Wang, G. X., & Chen, P. (2012). Screening of 42 medicinal plants for in vivo anthelmintic activity against *Dactylogyrus intermedius* (Monogenea) in goldfish (*Carassius auratus*). *Parasitology Research, 111*, 97–104.

Khan, M., Wassilew, S. W., Schmutterer, H., & Ascher, K. R. (1987). *Natural Pesticides from the Neem Tree and Other Tropical Plants*. GTZ: Eschborn, Germany, pp. 645–650.

Kirtikar, K. R., & Basu, B. D. (1933). *Medicinal Plants of India*. Pub. L. M. Basu: Allahabad, p. 3.

Kolkovski, S., & Kolkovski, J. (2011). Herbal medicine in Aquaculture. *International Aquafeed*, *14*(2), 28–31.

Koneri, R., Balraman, R., & Saraswati, C. D. (2006). Antiovulatory and abortifacient potential of the ethanolic extract of roots of *Momordica cymbalaria* fenzl in rats. *Indian Journal of Pharmacology*, *38*, 111.

Kragesteen, T. J., Simonsen, K., Visser, A. W., & Andersen, K. H. (2018). Identifying salmon lice transmission characteristics between Faroese salmon farms. *Aquaculture Environment Interactions*, *10*, 49–60.

Krishnan, K., Khanna, G. V., & Sahul Hameed, A. S. (2010). Antiviral activity of dasyscyphin C extracted from *Eclipta prostrata* against fish nodavirus. *Journal Antiviral and Antiretroviral*, *2*, 29–32.

Kulkarni, R. R., Pawar, P. V., Joseph, M. P., Akulwad, A. K., Sen, A., & Joshi, S. P. (2013). *Lavandula gibsoni* and *Plectranthus mollis* essential oils: Chemical analysis and insect control activities against *Aedes aegypti*, *Anopheles sfttephensi* and *Culex quinquefasciatus*. *Journal of Pest Science*, *86*(4), 713–718.

Kumar, A., Raman, R. P., Kumar, K., Pandey, P. K., Kumar, V., Mohanty, S., & Kumar, S. (2012). Antiparasitic efficacy of piperine against *Argulus* spp. on *Carassius auratus* (Linn. 1758): In-Vitro and In-Vivo Study. *Parasitology Research*, *111*, 2071–2076.

Lalumera, G. M., Calamari, D., Galli, P., Castiglioni, S., Crosa, G., & Fanelly, R. (2004). Preliminary investigation on the environmental occurrence and effects of antibiotics used in aquaculture in Italy. *Chemosphere*, *54*, 661–668.

Mitchell, A. J., & Hoffman, G. L. (1980). Important tapeworms of North American freshwater fishes. *Fish Disease Leaflet. Sr. No. 59, U.S. Fish and Wildlife Service*, p. 18.

Neeraja, B., & Narayana, G. (1998). Use of Neemol for cestode infected white leghorns to increase egg production. *Geobios*, *25*, 149–151.

Oketch-Rabah, H. A., Dossaji, S. F., Christensen, S. B., Frydenvang, K., Lemmich, E., Cornett, C., Olsen, C. E., Chen, M., Kharazm, A., & Theander, T. (1997). Antiprotozoal compounds from *Asparagus africanus*. *Journal of Natural Products*, *60*, 1017–1022.

Pachanawan, A., Phumkhachorn, P. and Rattanachaikunsopon, P. (2008). Potential of *Psidium guajava* supplemented fish diets in controlling *Aeromonas hydrophila* infection in tilapia (*Oreochromis niloticus*). *Journal of Bioscience and Bioengineering*, *5*, 419–424.

Parvati, S., & Kumar, V. J. (2002). Studies on chemical composition and utilization of wild edible vegetable Athalakkai (*Momordica tuberosa*). *Plant Foods for Human Nutrition*, *57*, 215.

Preena, P. G., Swaminathan, T. R., Kumar, V. J. R., & Singh, I. S. B. (2020). Antimicrobial resistance in aquaculture: A crisis for concern. *Biologia*, *75*, 1497–1517.

Purohit, S. S., & Mathur, S. K. (1999). *Drugs in Biotechnology—Fundamentals and Applications*. In: Purohit, S. S., (Ed.), Maximum Publishers: India.

Rahimi, N. N. M. K., Natrah, I., Loh, J. Y., Ranzil, F. K. E., Gina, M., Lim, S. H. E., Lai, K. S., & Chong, C. M. (2022). Phytocompounds as an alternative antimicrobial approach in aquaculture. *Antibiotics*, *11*(4), 469.

Raman, R. (2017). Applicability, feasibility and efficacy of phytotherapy in aquatic animal health management. *American Journal of Plant Sciences*, *8*(2), 257–287.

Raman, R. P. (2001). Herbal remedies for finfish and shellfish diseases and their commercialization. In: *National Seminar on Fishery Technologies and Their Commercialization, CIFE,* Mumbai, 11–12 January, p. 147.

Raman, R. P. (2004). Preliminary studies on antibacterial activity of medicinal plant extracts against some pathogenic bacteria of fish. In *National Symposium on Disease Problems in Aquaculture-Challenges, Approaches and Management, Central Institute of Fisheries Education, Mumbai*, p. 83.

Raman, R. P. (2007). Application of phytotherapy in finfish and shellfish health management: A Study. In: *Fisheries and Aquaculture: Strategic Outlook for Asia, 8th Asian Fisheries Forum, Asian Fisheries Society and AFSIB*, Kochi, India, 20–23 November 2007, p. 43.

Raman, R. P., Mukherjee, S. C., Rahman, M. K., & Kamble, S. (2004). Control of aquaculture diseases through traditional medicine systems of India. In: *National Conference on Traditional Knowledge Systems of India, Indian Institute of Technology (IIT),* Kharagpur, India, 9–11 January 2004.

Raman, R. P., & Rahman, M. K. (2002). *Phytotherapy in Fisheries*. World Ayurveda Congress, Book of Abstracts, Swadeshi Science Movement and Ministry of Health and Family Welfare, Government of India: Kochi, p. 42.

Rath, R. K. (2000). *Freshwater Aquaculture*. 2nd edn., Scientific Publishers: Jodhpur, India.

Rigos, G., Nengas, I., Alexis, M., & Troisi, G. M. (2004). Potential drug (oxytetracycline and oxolinic acid) pollution from Mediterranean sparid fish farms. *Aquatic Toxicology, 69*, 281–288.

Rodrigues, M. T., Alves, T. L. S., Soares, G. L. G., & Ritter, M. R. (2012). Plants used as antidiabetics in popular medicine in Rio Grande do Sul, southern Brazil. *Journal of Ethnopharmacology*, *139*, 155–163.

Romero Ormazabal, J. M., Feijoo, C. G., & NavarreteWallace, P. A. (2012). Antibiotics in aquaculture- use, abuse and alternatives. In: Carvalho, E. D., David, J. S., & Silva, R. J. (Eds.), *Health and Environment in Aquaculture*, p. 159.

Santos, L., Araujo, A. N., Fachini, A., Pena, A., Delerue-Matos, C., & Montenegrom, M. (2010). Ecotoxicological aspects related to the presence of pharmaceuticals in the aquatic environment. *Journal of Hazardous Materials*, *175*, 45–95.

Saurav, K., Raman, R. P., Pandey, P. K., Mohanty, S., Kumar, A., & Kumar, K. (2013). Effect of orally administered Azadirachtin on Non-Specific immune parameters of Goldfish *Carassius auratus* (Linn. 1758) and resistance against *Aeromonas Hydrophila. Fish & Shellfish Immunology*, *34*, 564–573.

Schar, D., Zhao, C., Wang, Y., Larsson, D. G., Gilbert, M., & Van Boeckel, T. P. (2021). Twenty-year trends in antimicrobial resistance from aquaculture and fisheries in Asia. *Nature Communications*, *12*, 5384.

Shalaby, A. M., Khattab, Y. A., & Abdel Rahman, A. M. (2006). Effects of garlic (*Allium sativum*) and chloramphenicol on growth performance, physiological parameter and

survival of Nile tilapia (*Oreochromis niloticus*). *Journal of Venomous Animals and Toxins Including Tropical Diseases*, *12*, 172–201.
Shaoqi, L. (1989). Main diseases and their control. In: *Integrated Fish Farming in China*. NACA Technical Manual 7. A World Food Day Publication of Network Centres in Asia and the Pacific: Bangkok, Thailand, p. 278.
Sharma, A., Deo, A. D., Ritesh kumar, S. T., Chanu, T. I., & Das, A. (2010). Effect of *Withania somnifera* (L. Dunal) root as a feed additive on immunological parameters and disease resistance to *Aeromonas hydrophila* in *Labeo rohita* (Hamilton) fingerlings. *Fish and Shellfish Immunology*, *29*, 508–512.
Sihag, R. C., & Sharma, P. (2012). Probiotics: The new eco-friendly alternative measures of disease control for sustainable aquaculture. *Journal of Fisheries and Aquatic Science*, *7*(2), 72–103.
Srinivasan, G., Babu, P. C., Sundara., & Murugeswari, V. (2001). Effect of neem products and insecticides on the egg parasitoids, *Trichogramma* spp. (Trichogrammatidae: Hymenoptera). *Pesticide Research Journal*, *13*, 250–253.
Sun, X., & Wang, Y. (1992). Study on measures for achieving high yield on Penaeid shrimp (*Penaeus chinensis*) culture. *Oceanologia et Limnologia Sinica*, *23*, 454–458.
Tiwari, B. N., Singh, K. P., Patra, P. K., Das, P. B., & Narain, R. (2006). Use of square banana stem technique for control of aquatic insects in nursery ponds (Abstract in Hindi). *National Seminar on Fisheries of North-Eastern States*, India, 6 December 2006, p. 84.
Vaidyaratnam, P. S. V. (1995). *Indian Medicinal Plants*. Arya Vaidya Sala, Kottakkal. Orient Longman Ltd.: Chennai.
Vasudeva, R. Y., Das, B. K., Jyotyrmayee, P., & Chakrabarti, R. (2006). Effect of *Achyranthes aspera* on the immunity and survival of *Labeo rohita* infected with *Aeromonas hydrophila*. *Fish and Shellfish Immunology*, *20*, 263–273.
Wang, W., Sun, J., Liu, C., & Xue, Z. (2017). Application of immunostimulants in aquaculture: Current knowledge and future perspectives. *Aquaculture Research*, *48*, 1–23.
Wei, L., & Musa, N. (2008). Inhibition of *Edward siella tarda* and other fish pathogens by *Allium sativum* L. (Alliaceae) extract. *Environmental Science, Agricultural and Food Sciences*, *3*, 692–696.
WHO (2004). *WHO Guidelines on Safety Monitoring of Herbal Medicines in Pharmacovigilance Systems*. World Health Organization: Geneva, Switzerland.
Wollenberger, L., Halling-Sorensen, B., & Kusk, K. O. (2000). Acute and chronic toxicity of veterinary antibiotics to *Daphnia magna*. *Chemosphere*, *40*, 723–730.
Woo, P. T. K., Bruno, D. W., & Lim, L. H. S. (2002). *Diseases and Disorders of Finfish in Cage Culture*. CABI.
Wu, Z. F., Zhu, B., Wang, Y., Lu, C., & Wang, G. X. (2011). In vivo evaluation of anthelmintic potential of medicinal plant extracts against *Dactylogyrus intermedius* (Monogenea) in goldfish (*Carassius auratus*). *Parasitology Research*, *108*, 1557–1563.
Xu, B., Ji, W., Zhang, P., Xu, H., & Shi, J. (1993). Comparision of antibacterial agents for control of pathogen in culture shrimp *Penaeus orientalis*. *Journal of Ocean University of China*, *23*, 43–57.
Yilmaz, E., Genc, M. A., Cek, S., Mazlum, Y., & Genc, E. (2006). Effects of orally administered *Ferula coskunii* (Apiaceae) on growth, body composition and histology of Common carp, *Cyprinus carpio*. *Journal of Animal and Veterinary Advances*, *5*, 1236–1238.

Zounkova, R., Klimesova, Z., Nepejchalova, L., & Hilscherova, B. L. K. (2011). Complex evaluation of ecotoxicity and genotoxicity of antimicrobials oxytetracycline and flumequine used in aquaculture. *Environmental Toxicology and Chemistry*, *30*, 1184–1189.

CHAPTER 8

Environmental DNA (eDNA) Surveillance: A Noninvasive Tool for Detecting Pathogens with Future Prospects

ASHISH SAHU[1,2], MAHENDER SINGH[1*], PRAMILA SAHADEVAN[2], and U. K. SARKAR[1]

[1]*National Bureau of Fish Genetic Resources, Lucknow, Uttar Pradesh, India*

[2]*Faculty of Fisheries, Kerala University of Fisheries and Ocean Studies, Cochin, Kerala, India*

*Corresponding author

ABSTRACT

To minimize adverse impacts on the economy and public health, the aquaculture sector must improve its surveillance technologies due to the growing prevalence of diverse pathogens. In aquaculture, environmental DNA (eDNA) techniques have emerged as an eco-friendly tool (noninvasive) for effective early detection and monitoring of pathogens and invasive taxa. Technological advancements have facilitated the easier detection of various aquatic organisms without causing environmental harm or impacting the animals themselves. In aquaculture, there are several pathogens detected by this molecular approach using Shotgun metagenomic sequencing. This chapter intends to help the researcher and farmers with the potential of eDNA-based detection of pathogens in aquaculture. Such knowledge can aid in predicting potential outbreaks before they occur, thereby enhancing disease management strategies and safeguarding the aquaculture industry's sustainability and productivity.

Advances in Freshwater Aquaculture: Technology for Sustainable Development. Tarang Kumar Shah, Madhu Sharma, Mohd. Ashraf Rather, & Avdhesh Kumar (Eds.)

DOI: 10.1201/9781779640857-8

8.1 INTRODUCTION

Aquaculture, often called "underwater farming," represents the most rapidly expanding sector in global food production (Mugimba et al., 2021; Sahu et al., 2022) and contributes to around 44% of total fish production. By 2013, it had already overtaken the production of beef and poultry (Sofia, 2018). This increased growth of production is achieved despite facing different problems in the aquaculture environment. Aquaculture production and development are declining dramatically due to several reasons. Among these, the emergence of pathogens is one of the most serious constraints, causing harm to farmers' livelihoods, loss of jobs, reduced income, and food insecurity (Lo et al., 2023). A total of 50% of fish production loss is because of diseases that are more severe in developing countries. China, one of the leading countries in aquaculture production, experiences a loss of 15% of its total fish production due to diseases. To address these challenges, environmental DNA (eDNA) methods have emerged as a promising tool for the detection and management of pathogens in aquaculture systems. This chapter highlights the application of eDNA methods in the context of aquaculture, focusing on their role in pathogen detection, monitoring, and management (Schadewell and Adams, 2021; Noji et al., 2021). eDNA is potentially a valuable tool for studying health-related issues in aquaculture sectors (Peters et al., 2018), including pharmaceuticals and other chemicals on cultured fish and shellfish. The microbial diversity of gut microflora could be also examined by using eDNA and next-generation sequencing (NGS) (D'Rose et al., 2019; Han et al., 2018).

8.2 WHAT IS EDNA?

eDNA technology is a powerful tool for species detection from water-containing cellular and extracellular DNA sloughed off all living animals (Bohmann et al., 2014; Sahu et al., 2023). eDNA tool has been successfully used to detect and monitor eukaryotic micro and macrobial communities and populations (Ficetola et al., 2008; Zamor et al., 2012). It is a widely useful approach for early monitoring systems as it allows for more accurate and standardized detection of species that are invasive, cryptic, inaccessible, endangered, rare, and of low abundance (Ficetola et al., 2008).

There have been several researchers that have validated eDNA as a monitoring method (Zamor et al., 2012) with some recent advances toward its application for pathogen detection in aquaculture systems (Gomes et al., 2017). However, before it can be considered as a systematic bio-monitoring tool, it is crucial to develop a cost-effective analytical approach to allow rapid processing of environmental samples on a daily basis. Furthermore, validation is required to establish the relationship between eDNA genotype data and biological abundance.

8.2.1 eDNA/ENVIRONMENTAL RNA (ERNA)

eDNA and eRNA are collectively known as eNA. Nucleic acids (DNA and/or RNA) isolated from environmental samples (soil, sediment, filtered water, air, etc.) and whole (micro-) organisms occurring in such samples (including pathogens, hyper parasites of tiny animals, and other more free-living microbes, and microbiomes). The scope of eNA also includes material released from animals but not sampled directly from them (e.g., feces and other excreta, blood, mucus, sloughed cells, pollen, gametes, etc.) (Sahu et al., 2023). Detection of parasites, pathogens, and other symbionts in environmental samples is one of the crucial sources of information about the presence or absence of organisms. eNA approach can detect as a proxy for infection of host animals in connected ecosystems, for pathogen monitoring and surveillance, and early warning systems for disease (Bohara et al., 2022)

8.2.2 PRINCIPLES OF eDNA IN AQUACULTURE

eDNA in aquaculture is based on the genetic material that cultured species release into their environment. These principles underpin the effective use of eDNA in aquaculture, providing a powerful tool for disease management, environmental monitoring, and sustainable aquaculture practices. Culture animals continuously release DNA through various processes, such as shedding skin cells, mucus, gametes, feces, and urine, into the surrounding environment, resulting in intraextracellular DNA being suspended in the water column and detectable for up to ~60 days (Strickler et al., 2015; Sahu et al., 2022). This eDNA serves as a molecular signature of the organisms present. eDNA collection is noninvasive, eliminating the need for physically interacting with aquatic species. This reduces stress on the animals and minimizes the risk of cross-contamination. The potential of this tool as a noninvasive inspection technique in the aquaculture sector was suggested by Asakura et al. (2014).

8.3 A GENERAL WORKFLOW OF EDNA APPROACH IN THE DETECTION OF PATHOGENS AND DISEASES FROM THE DIFFERENT AQUACULTURE SYSTEM

8.3.1 WATER SAMPLE COLLECTION

Collect water or sediment samples from the aquaculture system following the protocol (Goldberg et al., 2013) and ensuring proper handling and preservation to prevent DNA degradation. We can use three methods: (1) filtration method, (2) ethanol precipitation method, and (3) centrifugation and ultrafiltration. Negative controls should also be collected to check contamination. All equipment should be cleaned with 10% commercial bleach solution and sterilized using the autoclave or ultraviolet rays. One liter of water sample should be collected using sterile bottles for filtration. The ecological data should be also recorded at the sampling locations such as date, time, global positioning system of sampling locations, temperature, and power of hydrogen (pH). We should collect the infected fish samples also; if possible for further use.

8.3.2 WATER FILTRATION AND DNA EXTRACTION

All equipment for filtration and DNA extraction were cleaned with 10% bleach solution, and then with sterile nuclease-free water. After assembling the filtration unit, filter out the water through 0.22 μm pore sized, 47 mm diameter membrane (Millipore, MA, USA). After filtration, the filter membranes were removed using sterile forceps and put into 2.0 ml microtubes containing 95% ethanol, then stored at −20°C until the DNA extraction (Goldberg et al., 2013). The eDNA extraction and purification should carried out in a sterile room to avoid contamination, following the phenol–chloroform–isoamyl alcohol method, commercial kit, or liquid phase separation method. The DNA quality and quantification are checked by gel electrophoresis or nanodrop spectrophotometer. The extracted DNA samples are stored at –20°C until polymerase chain reaction (PCR) assay.

8.3.3 PCR AMPLIFICATION

Species-specific detection is based on the amplification and detection of short fragments of DNA from target organisms using PCR with species-specific primers (Bohmann et al., 2014). Amplify specific DNA regions or marker genes such as small subunit ribosomal RNA (16S for bacteria, 18S for eukaryotes) or mitochondrial (e.g., COI, 12S, Cytb) genes. The choice of primer sets depends on the pathogen being studied. To monitor possible contaminants, take negative PCR controls (ultrapure water) and sequence them in parallel with all samples.

8.3.4 LIBRARY PREPARATION AND DNA SEQUENCING

Prepare a DNA library from the PCR products. Sequence the DNA library using high-throughput sequencing platforms like Illumina and nanopore sequencing (Minion) according to guidelines. This step generates a massive amount of raw sequence data and then converts it into FASTQ format.

8.3.5 BIOINFORMATICS ANALYSIS

Use tools like FASTQC to assess the quality of raw sequence reads. Trim or filter low-quality reads using Trimmomatic or Cutadapt software (average quality score <30Q). Negative control samples that will fail DNA isolation, PCR amplification, or initial quality control checks after demultiplexing are not uploaded. Remove adapter sequences to ensure accurate down-stream analysis. Compare the sequences to reference databases for species identification and characterize the pathogens.

8.3.6 DATA INTERPRETATION, REPORTING, AND MANAGEMENT

Interpret the results to assess the health of the aquaculture system and make informed management decisions. Communicate findings to aqua-culture managers and stakeholders for necessary actions, such as disease control measures and biosecurity protocols.

8.3.7 MONITORING AND FOLLOW-UP

Implement ongoing monitoring to track pathogen dynamics and disease trends, adjusting management strategies as needed (Figure 8.1).

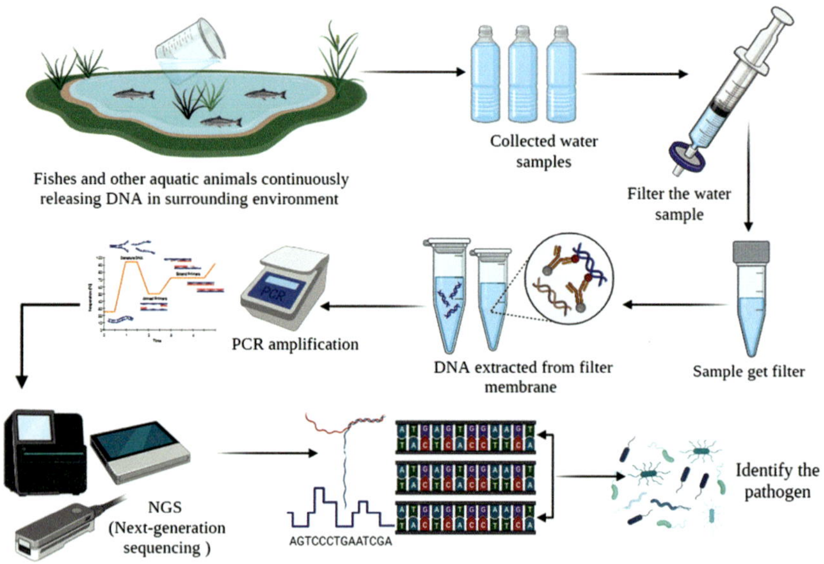

FIGURE 8.1 Schematic illustration of eDNA sampling from aquaculture pond. This figure represents the DNA fragments that are filtered out from the water, extracted the DNA, and detected the presence or absence of certain pathogens by using NGS.

8.4 APPLICATIONS OF EDNA IN AQUACULTURE AND FISH HEALTH MANAGEMENT

eDNA technology has several applications, helping in the early detection & monitoring of diseases & pathogens, and enhancing overall aquaculture and wild fish population management (Table 8.1).

1. **Early Disease Detection:** eDNA can be used to detect the presence or absence of pathogens at an early stage using specific genetic markers. This is crucial for preventing disease outbreaks and implementing timely interventions to minimize economic losses and enable proactive disease management. Common fish pathogens that can be monitored through eDNA include

TABLE 8.1 A List of Different Applications of eDNA for the Detection of Pathogens

Sl. No.	Application	Types of Water Body	Methodology	Conclusion	Citations
1.	Detection of koi herpesvirus (KHV) DNA	Tamagawa,_River, Japan	Virus concentration method and real-time PCR.	Help prevent outbreaks in carp farming	Haramoto et al., 2007
2.	Quantify Saprolegnia parasitica	Loue river and Tap water, France	Real-time quantitative PCR	Use of qPCR as an effective tool to monitor the dynamics of *S. parasitica.*	Rocchi et al., 2017
3.	Prediction of protozoan parasite outbreaks	Pond, Australia	Quantitative PCR (qPCR)	Quantify the abundance of parasitic protozoan that causes chilodonellosis in commercial freshwater finfish aquaculture	Gomes et al., 2017
4.	Monitoring biological threats to aquaculture, including pathogens in salmonid aquaculture.	Atlantic salmon farm, Norway, Ireland, and United Kingdom	Using amplicon-seq of the 18S SSU rDNA v9 region and Ion Torrent sequencing	Reliability of eDNA metabarcoding for monitoring aquaculture pathogens and harmful algae.	Peters et al., 2018
5.	Pathogen detection	Salmon Farm, British Columbia	Quantitative eDNA	Can assess pathogen information	Shea et al., 2020
6.	Detection and monitoring of fish ciliate parasite	Marine fish farm, Hong Kong	Droplet Digital PCR (ddPCR) Assay and NGS	Digital droplet PCR and eDNA are important new tool in clinical microbiology to quantify pathogens	Tsang et al., 2021
7.	Use of eDNA/RNA for pathogen and parasite detection, surveillance, and ecology.	Stream, Japan	Design and testing of a new set of primer-probe for detecting *E. multilocularis*	The detection rate of *E. multilocularis* using eDNA samples is low.	Mori et al., 2023
8.	Detection of multiple aquatic pathogens	Major waterways throughout Switzerland	Real-time quantitative PCR	eDNA can facilitate comprehensive disease monitoring schemes	Sieber et al., 2024

viruses (e.g., viral *Hemorrhagic septicemia*), bacteria (e.g., *Flavobacterium psychrophilum, Chilodonella hexasticha*), and several parasites (*Ichthyophthirius multifiliis*).

2. **Disease Surveillance and Biosecurity:** Continuous monitoring of culture ponds for the presence of fish pathogen DNA enables aqua culturists to assess the health status of their fish stock. Regular sampling and analysis of eDNA provide real-time insights into the health of aquatic systems, helping to identify changes in pathogen communities or environmental conditions. This proactive approach allows for the timely implementation of disease management strategies, reducing the need for invasive and potentially stressful fish sampling. eDNA can be used for biosecurity purposes in aquaculture facilities, which can help in identifying pathogens and potential contaminants that may pose a threat to the health of cultured fish. Early detection and containment can prevent the spread of diseases within the facility. eDNA serves as an early warning system for disease outbreaks. By detecting pathogen DNA before clinical signs appear, aqua culturists can implement timely interventions, reducing the impact of diseases on their stocks.
3. **Disease Management and Antibiotic Resistance Monitoring:** eDNA can be employed to evaluate the effectiveness of disease management, such as the use of vaccines or treatments, by monitoring changes in pathogen abundance or prevalence over time. The presence of antibiotic-resistance genes in the aquatic environment can be monitored through eDNA analysis. This is essential for assessing the development and spread of antibiotic resistance among fish pathogens, helping to guide antibiotic usage practices in aquaculture and breeding programs.
4. **Fish Stock Health Assessment:** eDNA analysis can be used to assess the overall health and well-being of fish stocks by monitoring stress-related biomarkers and other health indicators. It can provide insights into the physiological condition of fish, helping in timely interventions and adjustments to rearing conditions.
5. **Genetic Disease Screening:** Some genetic diseases and mutations in fish populations can be detected through eDNA analysis. Identifying these genetic issues early can inform breeding programs and prevent the spread of hereditary diseases within fish populations.

6. **Wild Fish Population Health:** eDNA can be used for monitoring the health of wild fish populations & stock, particularly in conservation efforts by assessing the presence of pathogens and diseases in the environment. Scientists can gauge the overall health and sustainability of wild fish populations.
7. **Epidemiological Studies:** eDNA technology can aid in conducting epidemiological studies to understand the spread and transmission dynamics of fish diseases within and between aquaculture facilities and natural water bodies.
8. **Environmental Impact Assessment:** By monitoring the presence of fish pathogens in the surrounding environment, aquaculture operations can assess their impact on the local ecosystem, aiding in responsible and sustainable management practices. Beyond pathogen detection, eDNA can reveal changes in microbial communities (Zamor et al., 2012), water quality, and environmental factors within aquaculture systems, aiding in overall environmental management.

8.5 CHALLENGES AND FUTURE DIRECTIONS OF EDNA IN AQUACULTURE AND FISH HEALTH MANAGEMENT

eDNA approach holds great promise for aquaculture and fish health management, but it also faces several challenges and offers numerous opportunities for future development in aquaculture and fish health management. We are addressing the limitations & challenges specific to eDNA analysis in aquaculture, including issues related to sample variability, detection limits, and the need for standardized protocols. Here are some of the following challenges for eDNA in this sector,

1. **Standardization and Methodology:** Standardization of eDNA sampling, extraction, and analysis protocols is essential to ensure the reliability and comparability of results across different studies and regions.
2. **Sensitivity and Specificity:** Improving the sensitivity and specificity of eDNA assays is crucial to accurately detect & identify species and pathogens, especially in complex aquatic environments with diverse microbial communities (Ficetola et al., 2008).
3. **Contamination and False Positives:** Reducing the risk of sample contamination and false positives is a significant challenge.

Proper sample collection, handling, and laboratory practices are essential to mitigate these issues.

4. **Quantification:** Quantifying the concentration of eDNA to estimate biomass and abundance remains a challenge. Developing robust quantitative eDNA methods is important for accurate stock assessment.
5. **Monitoring Dynamic Environments:** Aquatic environments are dynamic, with currents, temperature variations, and water quality changes. Incorporating these factors into eDNA analysis for real-time monitoring is challenging.
6. **Data Analysis and Bioinformatics:** In the eDNA study, NGS generates vast amounts of data, there is a need for advanced data analysis techniques and bioinformatics tools to extract meaningful information from eDNA datasets.

8.6 FUTURE PERSPECTIVES

The future perspectives of eDNA in aquaculture and fish health management are poised for remarkable advancements. With the ongoing evolution of molecular approaches and the integration of cutting-edge technologies, eDNA will play a pivotal role in shaping the aquaculture industry and fisheries management. Improved pathogen detection and monitoring will lead to early disease detection and rapid, targeted responses, significantly reducing economic losses in the aquaculture industry. The development of real-time eDNA monitoring systems will revolutionize the industry, allowing for continuous assessments of fish populations & aquatic environments, with immediate interventions when necessary. Precision aquaculture, driven by the eDNA tool, will become a reality, optimizing feeding, water quality, and environmental conditions in real-time to maximize production efficiency and minimize environmental impacts. Integrating eDNA data with other aquaculture management systems will provide a comprehensive view of operations, allowing for better decision-making. In-situ eDNA sampling and analysis with portable sequencing devices will eliminate the need for sample transportation. Continued advancements in DNA sequencing technologies, such as nanopore sequencing (mini ON), and the development of portable, field-deployable eDNA kits will make eDNA analysis more accessible and efficient. The emerging field of genomic epidemiology will enable the tracking of disease outbreaks, transmission routes, and the assessment of genetic diversity and resistance in fish populations, leading to more effective management strategies. Integrating eDNA analysis with microbiome studies can provide a more comprehensive understanding of aquatic ecosystems and fish health. This can help in differentiating between pathogenic and nonpathogenic microbes. Advanced data analysis techniques and artificial intelligence will enable predictive modeling for fish health, disease outbreaks, and environmental changes, empowering proactive management strategies, as well as assisting in analyzing large eDNA datasets. The development of standardized bioinformatics pipelines and data-sharing platforms will facilitate collaboration and data integration among researchers in the eDNA field. The integration of remote sensing technologies with eDNA can enhance monitoring capabilities by providing real-time, spatial information about aquatic environments. The analysis of biomarkers and stress indicators through eDNA technology will facilitate the early detection of health issues in fish populations, ensuring better overall fish welfare. Beyond aquaculture, eDNA will continue to be a vital tool for monitoring wild fish populations, assessing the impact of aquaculture on local ecosystems, and contributing to conservation efforts. As collaboration and data sharing become the norm across research, industry, and regulatory bodies, eDNA's potential and limitations will become better understood, and standardized practices will emerge. Governments and regulatory agencies will establish guidelines and regulations for eDNA technology's use in aquaculture and fisheries management to ensure ethical and responsible implementation. Public engagement and education will be instrumental in fostering awareness, understanding, and support for the technology. Development of regulations and policies for eDNA data collection and usage, particularly in the context of wild fish stock and environmental impact assessments, is necessary. Training programs for aquaculturists, researchers, and environmental professionals on the use of eDNA technology should be expanded to promote its effective adoption. Encouraging collaboration between biologisssts, geneticists, ecologists, and computer scientists can lead to innovative solutions and a deeper understanding of the potential and limitations of eDNA in aquaculture & fish health management.

Additionally, as technology advances and the industry embrace these innovations, the cost of eDNA analysis is likely to decrease, making it more accessible to a wider range of aquaculture operations and research projects. In the context of biosecurity, eDNA will continue to play a critical role in preventing disease outbreaks and the introduction of invasive species, safeguarding the aquaculture industry. Furthermore, eDNA will provide insights into the overall health and balance of aquatic ecosystems, allowing for more sustainable and responsible management practices. In summary, the future of eDNA technology in aquaculture and fish health holds the promise of more efficient and sustainable aquaculture practices, improved disease management, and a deeper understanding of aquatic ecosystems. As technology advances and the industry embraces these innovations, eDNA will play an increasingly pivotal role in ensuring the health and sustainability of fish populations and aquatic environments.

8.7 CONCLUSION

In conclusion, the eDNA approach is valuable and promising for the detection and management of fish health and pathogens in aquaculture systems. It allows for noninvasive, rapid, and cost-effective monitoring (Harper et al., 2019), which can lead to more effective disease prevention and control strategies in both aquaculture and wild fish stocks. By providing early detection, continuous monitoring, and data-driven management strategies, eDNA methods contribute to the sustainability and resilience of the aquaculture industry. This chapter has provided an overview of the principles, applications, challenges, and future directions of using eDNA for the detection and management of pathogens in aquaculture, emphasizing the potential for this technology to play a pivotal role in ensuring the health and productivity of aquaculture systems. Overall, the eDNA approach has the potential to revolutionize how we manage aquaculture and monitor fish health while addressing the existing challenges and pursuing these future directions.

KEYWORDS

- **pathogen**
- **environmental DNA**
- **aquaculture**
- **next-generation sequencing**
- **fish disease**

REFERENCES

Asakura, T., Sakata, K., Yoshida, S., Date, Y., & Kikuchi, J. (2014). Noninvasive analysis of metabolic changes following nutrient input into diverse fish species, as investigated by metabolic and microbial profiling approaches. *PeerJ*, *2*, e550.

Bohara, K., Yadav, A. K., & Joshi, P. (2022). Detection of fish pathogens in freshwater aquaculture using eDNA methods: A review. *Diversity*, *14*(12), 1015.

Bohmann, K., Evans, A., Gilbert, M. T. P., Carvalho, G. R., Creer, S., Knapp, M., Yu, D. W., & De Bruyn, M. (2014). Environmental DNA for wildlife biology and biodiversity monitoring. *Trends in Ecology & Evolution*, *29*(6), 358–367.

D'Rose, V., Johny, T. K., & Bhat, S. (2019). Comparative analysis of metagenomic DNA extraction methods from gut microbiota of zebrafish (Danio rerio) for downstream next-generation sequencing. *Journal of Applied Biology and Biotechnology*, *7*(1), 11–15.

Ficetola, G. F., Miaud, C., Pompanon, F., & Taberlet, P. (2008). Species detection using environmental DNA from water samples. *Biology Letters*, *4*(4), 423–425.

Goldberg, C. S., Sepulveda, A., Ray, A., Baumgardt, J., & Waits, L. P. (2013). Environmental DNA as a new method for early detection of New Zealand mudsnails (*Potamopyrgus antipodarum*). *Freshwater Science*, *32*(3), 792–800.

Gomes, G. B., Hutson, K. S., Domingos, J. A., Chung, C., Hayward, S., Miller, T. L., & Jerry, D. R. (2017). Use of environmental DNA (eDNA) and water quality data to predict protozoan parasites outbreaks in fish farms. *Aquaculture*, *479*, 467–473. https://doi.org/10.1016/j.aquaculture.2017.06.021

Han, Z., Sun, J., Lv, A., Sung, Y., Sun, X., Shi, H., & Xing, K. (2018). A modified method for genomic DNA extraction from the fish intestinal microflora. *AMB Express*, *8*, 1–8.

Haramoto, E., Kitajima, M., Katayama, H., & Ohgaki, S. (2007). Detection of koi herpesvirus DNA in river water in Japan. *Journal of Fish Diseases*, *30*(1), 59–61. https://doi.org/10.1111/j.1365-2761.2007.00778.x

Harper, L. R., Buxton, A. S., Rees, H. C., Bruce, K., Brys, R., Halfmaerten, D., & Hänfling, B. (2019). Prospects and challenges of environmental DNA (eDNA) monitoring in freshwater ponds. *Hydrobiologia*, *826*, 25–41.

Lo, L. S. H., Liu, X., Liu, H., Shao, M., Qian, P. Y., & Cheng, J. (2023). Aquaculture bacterial pathogen database: Pathogen monitoring and screening in coastal waters using environmental DNA. *Water Research X, 20*, 100194. https://doi.org/10.1016/j.wroa.2023.100194

Mori, K., Imamura, A., Hirayama, I., & Minamoto, T. (2023). Detection of Echinococcus multilocularis in repurposed environmental DNA samples from river water. *PeerJ, 11*, e15431.

Mugimba, K. K., Byarugaba, D. K., Mutoloki, S., Evensen, Ø., & Munang'andu, H. M. (2021). Challenges and solutions to viral diseases of finfish in marine aquaculture. *Pathogens, 10*(6), 673.

Noji, T., Wieczorek, D., Phelan, B., Liu, Y., Milke, L., Mercaldo-Allen, R., & Nash, B. (2021). Applications of environmental DNA data in support of aquaculture. In *Finally, I would Like to Express My Sincere Gratitude to the Colleagues Involved in the UJNR Aquaculture Panel for their Efforts in the Preparation and Organization of the Symposium. I would also Like to Deeply Thank the Editorial Board Members for Publishing the Proceedings.* Vol. 50, pp. 19–24.

Peters, L., Spatharis, S., Dario, M. A., Dwyer, T., Roca, I. J. T., & Kintner, A. (2018). Environmental DNA: A new low-cost monitoring tool for pathogens in salmonid aquaculture. *Frontiers in Microbiology 9*, 3009.

Rocchi, S., Tisserant, M., Valot, B., Laboissière, A., Frossard, V., & Reboux, G. (2017). Quantification of Saprolegnia parasitica in river water using real-time quantitative PCR: from massive fish mortality to tap drinking water. *International Journal of Environmental Health Research, 27*(1), 1–10.

Sahu, A., Abinaya, R., Amritha, P., Ansari, M. N., Abisha, C., Karthika, R., Kuma, V., & Masram, V. (2022a). Marine Fisheries Regulation Act (MFRA) and its amendments in India: An overview. In: *Research Trends in Fisheries and Aquatic Sciences.* Akinik Publications, Vol. 15, pp. 1–22. ISBN: 978-93- 5570-161-9 https://doi.org/10.22271/ed.book.1998.

Sahu, A., Kumar, N., Singh, C. P., & Singh, M. (2023a). Environmental DNA (eDNA): Powerful technique for biodiversity conservation. *Journal for Nature Conservation, 71*, 126325. https://doi.org/10.1016/j.jnc.2022.126325

Sahu, A., Singh, M., & Sarkar, U. K. (2023b). Environmental DNA (eDNA) advancement in conservation of Indian freshwater biodiversity with future perspectives. *Aquaculture Spectrum, 6*(9), 47–50.

Sahu, A., Singh, M., Kumar, N., Priyanka, & Naidu, B. C. (2022b). Environmental DNA (eDNA): A novel revolutionary tool and its application in biodiversity assessment. In: *National Symposium on Fisheries and Aquaculture for Livelihood and Nutritional Security*, ICAR-DCFR: Bhimtal, p. 65.

Schadewell, Y., & Adams, C. I. (2021). Forensics meets ecology–environmental DNA offers new capabilities for marine ecosystem and fisheries research. *Frontiers in Marine Science, 8*, 668822. https://doi.org/10.3389/fmars.2021.668822

Shea, D., Bateman, A., Li, S., Tabata, A., Schulze, A., Mordecai, G., & Krkošek, M. (2020). Environmental DNA from multiple pathogens is elevated near active Atlantic salmon farms. *Proceedings of the Royal Society B, 287*(1937), 20202010.

Sieber, N., King, A., Krieg, R., Zenker, A., Vorburger, C., & Hartikainen, H. (2024). Large-scale eDNA monitoring of multiple aquatic pathogens as a tool to provide risk maps for wildlife diseases. *Environmental DNA, 6*(1), e427.

Sofia, F. (2018). *The State of World Fisheries and Aquaculture 2018-Meeting the Sustainable Development Goals*. Fisheries and Aquaculture Department, Food and Agriculture Organization of the United Nations: Rome.

Strickler, K. M., Fremier, A. K., & Goldberg, C. S. (2015). Quantifying effects of UV-B, temperature, and pH on eDNA degradation in aquatic microcosms. *Biological Conservation*, *183*, 85–92. https://doi.org/10.1016/j.biocon.2014.11.038

Tsang, H. H., Domingos, J. A., Westaway, J. A., Kam, M. H., Huerlimann, R., & Bastos Gomes, G. (2021). Digital droplet PCR-based environmental DNA tool for monitoring *Cryptocaryon irritans* in a marine fish farm from Hong Kong. *Diversity*, *13*(8), 350. https://doi.org/10.3390/d13080350

Zamor, R. M., Glenn, K. L., & Hambright, K. D. (2012). Incorporating molecular tools into routine HAB monitoring programs: Using qPCR to track invasive Prymnesium. *Harmful Algae*, *15*, 1–7.

CHAPTER 9

Role of Probiotics in Aquaculture

N. UDAY KIRAN[1], SHIWAM DUBEY[1*], S. VIJAYAKUMAR[2], KOMMU MADHU[3], and A. VINOD KUMAR[3]

[1]*Karnataka Veterinary, Animal and Fisheries Sciences University, Bidar, College of Fisheries, Mangaluru, Karnataka, India*

[2]*Fisheries Research and Information Centre, Vijayapur, Karnataka, India*

[3]*P. V. Narasimha Rao Veterinary University, College of Fishery Science, Pebbair, Telangana, India*

*Corresponding author

ABSTRACT

A major part of the Indian economy is the fishing sector. It generates jobs, boosts exports, ensures food and nutrition security, and increases national income. About 30 million people in India depend on the fishing industry, which is known as the "Sunrise Sector" and especially supports the weaker and vulnerable communities. The primary industry to meet the rising demand for seafood in the future is predicted to be aquaculture, a significant source of food supply. Nevertheless, the aquaculture industry is facing significant challenges as a result of infectious diseases causing detrimental impacts. The emergence of resistant bacterial strains is attributed to the widespread application of antimicrobial drugs, pesticides, and disinfectants for disease prevention and growth enhancement in fish farming. Consequently, there is a growing interest in exploring probiotics as an alternative approach for sustainable aquaculture practices. Probiotics are bacteria that have an important role in aquaculture promoting the health

Advances in Freshwater Aquaculture: Technology for Sustainable Development. Tarang Kumar Shah, Madhu Sharma, Mohd. Ashraf Rather, & Avdhesh Kumar (Eds.)

DOI: 10.1201/9781779640857-9

of the host organism and aiding in protection against harmful bacterial infections. This chapter summarizes the role of probiotics in aquaculture.

9.1 INTRODUCTION

The population of humans is expected to grow to 9.7 billion by 2050 and 10.4 billion by 2100, according to UN estimates (United Nations, 2022). Food security and sustainability have become issues due to population growth and the ensuing climate change. Food with superior protein, vitamin, and mineral values can be obtained through aquaculture, according to some presentations (Maulu et al., 2021). The Food and Agriculture Organization in 2020 reported that the global fish production was estimated to be 179 million tons, valued at 401 billion USD. Of this total, aquaculture accounted for 46%, producing 82.1 million tons, of which the main species are finfish (53.2 million tons), mollusks (17.7 million tons), and crustaceans (9.4 million tons).

Total fisheries and aquaculture production in 2020 was 214 Mt, in which capture fisheries were around 178 Mt and aquaculture production was around 87.5 Mt, aquaculture has shown an increasing trend of 2.5% than the previous year. However, aquaculture intensification leads to overcrowding and degrading of water quality, facilitating the spread of pathogens and increasing disease outbreaks and mortality (Bondad-Reantaso et al., 2005). In order to achieve the sustainable development of aquaculture, namely, disease prevention, improving resistance to pathogens, increasing growth performance, immunomodulation, and maintenance of health of cultured fish has become the main concern. To overcome these problems, antibiotics were used previously as immunostimulants. The aquatic environment is intimately linked with human life because of the widespread consumption of fish and shrimp. When antibiotics are used extensively and continuously in aquaculture environment, there may be a number of risks to human health as well as aquatic organism health when fish and shrimp are consumed. Antibiotics used in aquaculture, whether for preventative or therapeutic reasons, frequently accumulate in aquatic animals' tissues. Antimicrobial drug residues in edible tissues can result in adverse reactions, toxicity, altered gut microbial ecology, and drug resistance (Figure 9.1). Probiotics in aquaculture provides an approach which is economically profitable and ecologically safe.

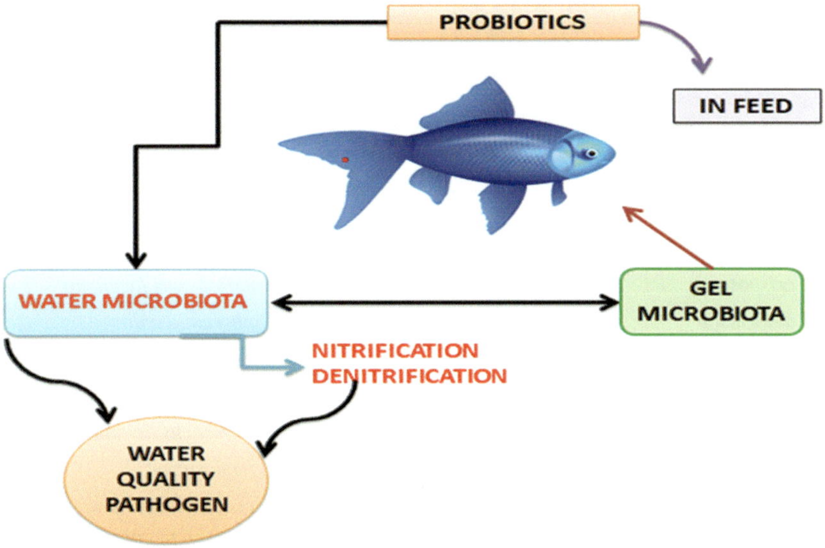

FIGURE 9.1 Role of probiotics in aquaculture.

9.2 WHAT IS PROBIOTIC?

The term probiotic is derived from the Greek word pro and bios, meaning of life. Probiotics in aquaculture are defined as a live microbial feed supplement that includes bacteria and nonbacteria that are applied via as a feed supplement or to the rearing water (Fuller and Fuller, 1992). These are live microbes that can confer a health benefit to host by improve the gastrointestinal microbes balancing and growth performance.

9.2.1 PROBIOTICS USED IN AQUACULTURE

Over three decades have passed since probiotics were initially used in aquaculture species, although significant attention was not provided until the early 2000s. because of its advantages and its ease of use in aquaculture. Probiotics are beneficial to the host by improving growth (Kumar et al., 2006), improving disease resistance (Irianto and Austin, 2009), enhancing physiological and immune responses (Dawood et al., 2016), improving water quality (Lalloo et al., 2007), and reducing the use of chemicals and antibiotics. The commonly used probiotics belong to the genus *Lactobacilli*, *Bifidobacteria*, *Bacillus*, and some other species (Iqbal et al., 2014) (Table 9.1).

TABLE 9.1 Probiotics Used in Aquaculture

Lactobacilli	Bifidobacteria	Other species
L. acidophilus	*B. adolescentis*	*Bacillus subtilis*
L. casei	*B. animalis*	*Enterococcus faecalis*
L. crispatus	*B. bifidum*	*Enterococcus faeciumn*
L. delbruckit	*B. breve*	*Escherichia coli*
L. gallinarum	*B. infantis*	*Lactococcus lactis*
L. gasseri	*B. lactis*	*Leucnostoc mesenteroides*
L. Johnsonii	*B. longum*	*Pediococcus acidilactici*
L. paracaset		*P. pentosaceus*
L. plantarum		*Saccharomyces boulardit*
L. reuteri		*Sporolactobacillus inulinus*
L. rhamosus		*Streptococcus thermophilis*

9.3 SELECTION CRITERIA FOR PROBIOTICS

- Nonpathogenicity
- Adherence to gut epithelial cells
- Colonization of GI tract
- Resistance to gastric juice
- Resistance to low pH
- Resistant to antibiotics
- Production of antimicrobial substances
- Beneficial effects on health
- Stimulation of immune response
- Viability in processing and storage
- Safety Properties in food and clinical use

9.4 TYPES OF PROBIOTICS

Based on the mode of application probiotics can be divided into two broad categories (Table 9.2):

TABLE 9.2 Different Species of Probiotics Used in Aquaculture and Their Beneficial Effects

Probiotic Candidates	Host Species	Dosage Duration	Beneficial Effects	References
Micrococcus luteus, Pseudomonas species	Nile tilapia	Mixture of both 10^{10} cells/kg diet 90 days	*M. luteus* ↑ FBW and WG, and ↓ FCR Pseudomonas sp. ↓ FBW, WG, SGR, PER, and SR%	Abd El-Rhman et al. (2009)
B. pumilus	Nile tilapia	10^9 or 10^{15} CFU/kg diet 1–2 months	↑SR% at 10^{15} CFU/kg diet in 1 month No change in growth	Aly et al. (2009)
B. subtilis E20	Grouper	10^7, 10^9, and 10^{11} CFU/kg feed 28 d	↑SR% with ↑probiotic concentration in the diet Accelerated larval development ↑WG and FE with ↑B. subtilis concentration	Liu et al. (2012)
L. acidophillus	Nile tilapia	10^9 CFU/kg 15 d	↑SR% No differences in FBW and WG	Villamil et al. (2014)
S. cerevisiae	Nile tilapia	0.25, 0.50, 1.0, 2.0, and 5.0 g/kg 84 d	Decreased mortality with increased yeast supplementation	Abdel-Tawwab et al. (2008)

1. **Gut probiotics:** To enhance the healthy microbial flora associated with the gut, probiotics are administered orally with food.
2. **Water probiotics:** Water probiotics that are added to the water to improve the water quality parameters by improving the water-associated microbial flora.

9.4.1 *MECHANISM OF ACTION*

- Competitive exclusion.
- Production of inhibitory compounds.
- Competition for nutrients.
- Quorum sensing (QS) manipulation.
- Immunomodulation.
- Improving water quality.

1. **Competitive exclusion:** For the development of a disease, many harmful bacteria must adhere to the gastrointestinal tract mucosa of their host. Competition for attachment sites, or "competitive exclusion," is a key mode of action in probiotic microorganisms. One ideal criterion for choosing probiotics is the ability of bacteria to colonize the gut, stick to the epithelial surface, and so prevent pathogen adherence. Pathogenic bacteria are eliminated from the stomach by probiotic bacteria because they compete with them and leave no room for attachment.
2. **Production of inhibitory substances:** Probiotic bacteria generate a variety of compounds including bacteriocins hydrogen peroxide siderophores, lysozymes, and proteases, which has a bactericidal or bacteriostatic effect on various microbial populations. Furthermore, certain bacteria can lower the pH in the gastrointestinal lumen by producing organic acid and volatile fatty acids (such as lactic, acetic, butyric, and propionic acids), which inhibits the growth of opportunistic pathogenic germs (Tinh et al., 2008).

   Many microorganisms, including the known probiotic group lactic acid bacteria, consume the nutrients that are essential for the growth of some pathogens. The existence of any microbial population depends on its ability to compete for chemicals and available energy with the other microbes in the same environment (Verschuere et al., 2000).

3. **Competition for nutrients:** Numerous microbes, such as the well-known probiotic lactic acid bacteria group, eat the nutrients required for some pathogens to grow. Any microbial population's capacity to contend with other organisms in the same environment for available resources and compounds is essential to its survival (Verschuere et al., 2000).
4. **Disruption of QS:** QS is a process of cell communication that allows bacteria to share information about cell density and adjust gene expression accordingly. Probiotic bacteria interfere with the pathogenic bacteria and inhibits the gene expression that causes pathogenicity. This system is used by a large number of bacteria to communicate and control a wide range of physiological processes. In aquaculture, QS disruption is seen to be a viable anti-infective strategy (Defoirdt et al., 2004).
5. **Immunomodulation:** Probiotics work by stimulating the host's immune system, which includes boosting pro-inflammatory cytokines on immune cell activity, raising leukocyte phagocytic activity, of leukocytes raising the amounts of antibodies, acid phosphatase, lysozymes, and antimicrobial peptides thereby increasing the immune performance and disease resistance (Pirarat et al., 2006).
6. **Interference of QS:** Nutrients necessary for the growth of some pathogens are consumed by a variety of microorganisms, including the well-known probiotic group, lactic acid bacteria. The ability of any microbial population to compete for chemicals and available energy with other microorganisms in the same environment determines its survival (Verschuere et al., 2000).
7. **Improving water quality:** The use of Gram-positive bacteria, like Bacillus species, can help to improve the water system's quality because they are more effective at converting organic matter into carbon dioxide than Gram-negative bacteria, which turn more organic substance in bacterial bio-mass (Balcazar et al., 2006). Significant algicidal effects are also exhibited by several probiotic bacteria, especially on a variety of microalgae species. Applying

nitrifying cultures to the fish habitat helps remove the toxicity of ammonia and nitrite. Furthermore, probiotics are advantageous because they can change the quality of the water and change the mix of microbial species in it.

8. **Amelioration of the effects of stress:** Amelioration of the effects of stress one way to think of stress is as a chemical or physical agent that triggers reactions that have the potential to cause illness and even death. Aquatic creatures' physiology and behavior may be negatively impacted by any alteration in the water's properties. Fish may be negatively impacted by a variety of stresses, such as temperature, nutritional, high stocking density, anoxia, hypoxia, pesticides, and poisons. Adding probiotic bacteria to water or feed can help avoid stressful situations by strengthening the immune system and lessening the negative impacts of certain stressors.

9.5 SIDE EFFECTS AND MISUSE

Probiotics are generally considered safe and effective (Boyle et al., 2006). The possibility that these live organisms could leave the gastrointestinal system and lead to systemic illnesses is one possible worry about probiotics (Snydman, 2008). Although it has not been seen yet, probiotics may also increase the possibility of pathogenic bacteria acquiring antibiotic resistance from probiotic strains (Martin et al., 2013). Systemic infections and financial losses in fish farms can be caused by the importation of probiotics into the aquaculture sector, changes in intestinal microbiota, new illnesses, and bacterial DNA mutagenesis or recombination (Ringo et al., 2010).

9.6 CONCLUSION

Approximately 90% of global production comes from aquaculture, which is one of the industries with the fastest growth rates in the world. Probiotics are a possible substitute for antibiotics in fish aquaculture that can lessen the detrimental effects of pathogen outbreaks, the financial losses resulting from specimen mortality, and the need for antibiotics to combat bacterial infections. These elements will contribute to the fish aquaculture industry's increased environmental friendliness. Since humans and other mammals have been the subjects of most studies and applications of probiotics explored in fish aquaculture, it is imperative to develop novel probiotics specifically tailored for fish use.

KEYWORDS

- **aquaculture**
- **probiotics**
- **antibiotics**
- **fisheries**
- **growth**

REFERENCES

Abd El-Rhman, A. M., Khattab, Y. A., & Shalaby, A. M. (2009). Micrococcus luteus and pseudomonas species as probiotics for promoting the growth performance and health of Nile tilapia, *Oreochromis niloticus*. *Fish & Shellfish Immunology*, *27*(2), 175–180.

Abdel-Tawwab, M., Abdel-Rahman, A. M., & Ismael, N. E. M. (2008). Evaluation of commercial live bakers' yeast, Saccharomyces cerevisiae as a growth and immunity promoter for Fry Nile tilapia, Oreochromis niloticus (L.) challenged in situ with Aeromonas hydrophila. *Aquaculture*, *280*(1), 185–189.

Aly, S. M., Abd-El-Rahman, A. M., John, G., & Mohamed, M. F. (2008). Characterization of some bacteria isolated from Oreochromis niloticus and their potential use as probiotics. *Aquaculture, 277*(1), 1–6.

Austin, B., & Brunt, J. W. (2009). The use of probiotics in aquaculture. *Aquaculture Microbiology and Biotechnology*, *1*, 185–207.

Bondad-Reantaso, M. G., Subasinghe, R. P., Arthur, J. R., Ogawa, K., Chinabut, S., Adlard, R. Tan, Z., & Shariff, M. (2005). Disease and health management in Asian aquaculture. *Veterinary Parasitology*, *132*(3–4), 249–272.

Boyle, R. J., Robins-Browne, R. M, & Tang, M. L. K. (2006). Probiotic use in clinical practice: What are the risks? review articles. *The American Journal of Clinical Nutrition*, *83*, 1256–1264.

Dawood, M. A., Koshio, S., Ishikawa, M., El-Sabagh, M., Esteban, M. A., & Zaineldin, A. I. (2016). Probiotics as an environment-friendly approach to enhance red sea bream, pagrus major growth, immune response and oxidative status. *Fish & Shellfish Immunology*, *57*, 170–178

Defoirdt, T., Boon, N., Bossier, P., & Verstraete, W. (2004). Disruption of bacterial quorum sensing: An unexplored strategy to fight infections in aquaculture. *Aquaculture*, *240*(1–4), 69–88.

El-Saadony, M. T., Alagawany, M., Patra, A. K., Kar, I., Tiwari, R., Dawood, M. A., Dhama, K., & Abdel-Latif, H. M. (2021). The functionality of probiotics in aquaculture: An overview. *Fish & Shellfish Immunology*, *117*, 36–52.

FAO. (2020). The state of food and agriculture 2020. In: *Overcoming Water Challenges in Agriculture*. Rome. https://doi.org/10.4060/cb1447en

FAO. (2022). The state of world fisheries and aquaculture 2022. *Towards Blue Transformation*. FAO: Rome. https://doi.org/10.4060/cc0461en

Fuller, R., & Fuller, R. (1992). History and development of probiotics. In: *Probiotics: The Scientific Basis*, pp.1–8.

Hai, N. V. (2015). The use of probiotics in aquaculture. *Journal of Applied Microbiology*, *119(*4), 917–935.

Iqbal, M. Z., Qadir, M. I., Hussain, T., Janbaz, K. H., Khan, Y. H., & Ahmad, B. (2014). Probiotics and their beneficial effects against various diseases. *Pakistan Journal of Pharmaceutical Sciences*, *27*(2), 405–415.

Irianto, A., & Austin, B. (2002). probiotics in aquaculture. *Journal of fish diseases*, *25*(11), 633–642.

Kumar, R., Mukherjee, S. C., Prasad, K. P., & Pal, A. K. (2006). Evaluation of bacillus subtilis as a probiotic to Indian major carp labeo rohita (ham.). *Aquaculture Research*, *37*(12), 1215–1221.

Lalloo, R., Ramchuran, S., Ramduth, D., Görgens, J., & Gardiner, N. (2007). Isolation and selection of Bacillus spp. as potential biological agents for enhancement of water quality in culture of ornamental fish. *Journal of Applied Microbiology*, *103*(5), 1471–1479.

Liu, C.-H., Chiu, C.-H., Wang, S.-W., & Cheng, W. (2012). Dietary administration of the probiotic, Bacillus subtilis E20, enhances the growth, innate immune responses, and disease resistance of the grouper, *Epinephelus coioides*, *Fish Shellfish Immunol*, *33*(4), 699–706.

Martin, H., Floch, M. D, & Macg, A. (2013). Probiotic safety and risk factors. *Journal of Clinical Gastroenterology*, *47*(5), 375–376.

Maulu, S., Hasimuna, O. J., Haambiya, L. H., Monde, C., Musuka, C. G., Makorwa, T. H., Munganga, B. P., Phiri, K. J., & Nsekanabo, J. D. (2021). Climate change effects on aquaculture production: Sustainability implications, mitigation, and adaptations. *Frontiers in Sustainable Food Systems*, *5,* 70.

Pirarat, N., Kobayashi, T., Katagiri, T., Maita, M., & Endo, M. (2006). Protective effects and mechanisms of a probiotic bacterium *Lactobacillus rhamnosus* against experimental Edwardsiella tarda infection in tilapia (*Oreochromis niloticus*). *Veterinary Immunology and Immunopathology*, *113*(3–4), 339–347.

Ringø, E., Olsen, R. E, Gifstad, T., Dalmo, R. A, Amlund, H., Hemre, G., & Bakke, A. M. (2010). Probiotics in aquaculture: A review. *Aquaculture Nutrition*, *16*, 117–136.

Snydman, D. R. (2008). The safety of probiotics. *Clinical Infectious Diseases*, *46*(2), 1041–1011.

Tinh, N. T. N., Dierckens, K., Sorgeloos, P., & Bossier, P. (2008). A review of the functionality of probiotics in the larviculture food chain. *Marine Biotechnology*, *10*, 1–12.

Tuan, T. N., Duc, P. M., & Hatai, K. (2013). Overview of the use of probiotics in aquaculture. *International Journal of Research in Fisheries and Aquaculture*, *3*(3), 89–97.

United Nations. (2022). *World Population Prospects 2022 Summary of Results*. United Nations: New York, NY, USA.

Verschuere, L., Rombaut, G., Sorgeloos, P., & Verstraete, W. (2000). Probiotic bacteria as biological control agents in aquaculture. *Microbiology and Molecular Biology Reviews*, *64*(4), 655–671.

Villamil, L., Reyes, C., & Martínez-silva, M. A. (2014). In vivo and in vitro assessment of lactobacillus acidophilus as probiotic for tilapia (*Oreochromis niloticus*, perciformes: Cichlidae) culture improvement. *Aquaculture Research, 45*(7), 1116–1125.

CHAPTER 10

Impact of Climate Change on Inland Fisheries in India

SHIWAM DUBEY[1*], GOODLY SAMAL[2], AITWAR VAIJNATH[1], BINAL KHALASI[1], LAXMI PRASAD[3], and JYOTI SAROJ[3]

[1]*Karnataka Veterinary, Animal and Fisheries Sciences University, Bidar, College of Fisheries, Mangaluru, Karnataka, India*

[2]*Dr. Rajendra Prasad Central Agricultural University, Pusa, College of Fisheries, Dholi, Bihar, India*

[3]*Acharya Narendra Deva University of Agriculture and Technology, Kumarganj, Ayodhya, Uttar Pradesh, India*

*Corresponding author

ABSTRACT

Fisheries and aquaculture are two of the most popular activities in coastal areas worldwide. Additionally, in the context of climate change, both activities are high risk because they are susceptible to environmental changes that may threaten the socioeconomic sustainability of fishing communities that depend on fish for both food security and income generation. Aquaculture has become the food production industry with the fastest growth rate in the world due to its continued significant production expansion. However, the sector's sustainability is at threat because of the anticipated effects of climate change, which are both a present and a future reality. In this chapter, we provide details about how climate change may affect aquaculture production and how that may affect the sustainability of the industry.

Advances in Freshwater Aquaculture: Technology for Sustainable Development. Tarang Kumar Shah, Madhu Sharma, Mohd. Ashraf Rather, & Avdhesh Kumar (Eds.)

DOI: 10.1201/9781779640857-10

10.1 INTRODUCTION

According to recent statistics, over 36,248 fish species have been identified globally, comprising roughly 18,336 freshwater species, with another 149 species expected to be added in 2022. Fishes are uniquely variable in morphology, biology, numbers, and size as ecological conditions influence the distribution of fishes in a particular region. Fish live in every type of water on the planet, from the deepest lake to hot soda water that can reach 44° in the Antarctic. Fish are important because they are a key source of protein that is both highly valuable and readily available, and they play a vital role in the development of the nation. Apart from being a rich source of highly nutritious protein and nutrients which are essential for the human body. In the last 40 years, the global marine catch has more than quadrupled, rising from 18.5 million tons in 1950 to 178.5 million tons in 2018, with China at the topmost (FAO, 2020).

10.2 INDIAN MARINE FISHERIES OVERVIEW AND FISH PRODUCTION SCENARIO

India has a 2.02 million km^2 exclusive economic zone and 8129 km of coast-line. Area of the continental shelf: 0.51 million km^2. On the eastern side, it is bound by the Bay of Bengal, and on the western side by the Arabian Sea. In the south of India, the Andaman and Nicobar Islands lie at a distance of 1200–1800 km from the east coast of the mainland and the Lakshadweep Islands at a distance of 200–400 km from the west coast (CMFRI, 2014). This observance provides nutritional security, income, and employment to the people of the country. The rapid increase in fish production has largely resulted from increased inland fishing. Most people in India still depend on fishing as a source of income, and it was initially a capture-based economy. India produced more than 162.48 lakh tons of fish in 2021–2022, of which marine fisheries accounted for 41.27 lakh tons, and inland fisheries accounted for 121.121 lakh tons. Out of which 18.7% was obtained from marine fisheries and 7.76% from inland fisheries (FAO, 2022).

10.2.1 ENVIRONMENT TRENDS

The tropical monsoon has an impact on the climate. From October through February, there are strong northeast winds, and from May through

October, there are predominant southwest winds. Cyclones hit the shores, particularly in the Bay of Bengal, when the monsoon winds move. Changes in the important oceanic weather systems such as sea surface temperature, pH, salinity, El Niño Southern Oscillation, precipitation, sea level, frequency and intensity of cyclones, and droughts are becoming evident as a result of climate change (Prasanna Kumar, 2010).

10.2.2 IMPACT ON BIOLOGICAL PROCESSES

It is expected that these changing physical forces would have a variety of effects on tropical coastal fish and there are indications that some of these are already occurring. Under the "National Innovation on Climate Resilient Agriculture" project, which is funded and managed by the Indian Council of Agricultural Research, New Delhi, and the Central Marine Fisheries Research Institute in India, is evaluating the effects of climate change on marine organisms, especially fish. The Network Project on Climate Change, which came before this one, has contributed to our understanding of the effects, vulnerabilities, and adaptations of the food-producing sectors, particularly fisheries, to climate change.

Marine organism species respond to climate change in different ways. The habitats, geographic affinities, and biological characteristics of the species all influence the impact and adaptation. Animals frequently adapt to climate change by expanding their geographic ranges, with many species doing so to higher latitudes. Due to their ability to move both horizontally and vertically, the majority of marine fish species are able to either adapt to new environments or use directed movement to protect themselves from adverse weather conditions. The findings that have come from the outcomes of these projects include changes in fertility, growth, spawning phenology, age/size at maturity, and growth.

10.2.3 IMPACT OF CHANGING CLIMATE ON THE ECOSYSTEM

Climate change is a serious issue with far-reaching repercussions; India's interior fisheries are particularly affected. A vital part of India's socioeconomic structure, inland fisheries support millions of people's livelihoods and provide a sizable contribution to the country's food security. Due to climate change, impacts on these ecosystems are occurring at a rapid pace and are causing a variety of problems for populated areas. India is particularly vulnerable to changes in hydrological patterns, temperature regimes, and biological dynamics brought on by climate change due to its different climatic zones and enormous network of rivers, lakes, and reservoirs. Fish populations, habitats, and the communities that depend on them are all disrupted by these changes, which impact aquatic life. Additionally, it discusses the economic and social implications of climate change on India's inland fishing industry, emphasizing the critical need for mitigation and adaptation measures (Table 10.1).

10.2.4 ALTERED TEMPERATURE REGIMES

Understanding the intricate physiological, ecological, and biological processes that underpin the relationships between climate and biodiversity is essential to estimating the ecological effects of climate change (Bellard et al., 2012). Ecological patterns responding to climate change, such as changes in species phonologies, distributions of body sizes, and rates of ecosystem function, have been largely revealed by long-term records. By using "climate envelope modeling," for example, species distributions are predicted based on physiological thresholds (Parmesan, 2006). Over the past few decades, experiments have been utilized more and more to clarify fundamental mechanisms and comprehend the implications of climate change. The majority of experimental research has looked into the possible effects of rising mean climate conditions. It is now clear that variations in the occurrence, length, and severity of extreme weather events, such as heatwaves and dry spells, are layered with trend effects. We need a better understanding of the implications of plausible future weather scenarios that include increases in mean temperatures as well as changes in the intensity, frequency, and amplitude of extreme events to forecast the effects of future climate changes on biota (Thompson et al., 2013). Significant changes in Indian rivers, lakes, and other freshwater bodies' temperature regimes as a result of climate change have far-reaching effects on aquatic ecosystems, particularly fish populations.

10.2.5 DISRUPTION OF THERMAL PREFERENCES

Temperature affects cellular functions that take place in an organism. If the temperature changes, each fish has to either adapt internally (a

TABLE 10.1 Evidence for Climate Change Impacts on Indian Marine Fish and Fisheries

Sr. No	Impact	Species	Response	Cause	Source
1	Distribution	*Sardinella longiceps, Rastrelliger kanagurta*	Extension of the distributional boundary both vertically and horizontally	Seawater warming	Vivekanandan (2011)
2	Phenology	*Nemipterus japonicas, Nemipterus mesoprion*	Peak spawning shifting toward cooler months	Seawater warming	Vivekanandan (2011)
3	Maturity	*Fenneropenaeus indicus, Metapenaeus monoceros, Metapenaeus dobsoni*	Reduction in length at first maturity	Seawater warming, overfishing	CMFRI (2014)
4	Fecundity	*Metapenaeus monoceros, Metapenaeus dobsoni*	Reduction in number of eggs	Precocious maturity	CMFRI (2014)
5	Growth	*Sardinella longiceps, Nemipterus japonicas, Rastrelliger kanagurta*	Faster growth rate	Seawater warming	Vivekanandan (2013)
6	Length in fishery	*Nemipterus japonicus*	Reduction in mean length in fishery	Seawater warming, overfishing	CMFRI (2014)
7	Fishing down food web	All species	Reduction in mean trophic level of fishery	Seawater warming	Vivekanandan (2013)
8	Coral bleaching	Corals	Increase in frequency and intensity of bleaching	Seawater warming, acidification	Vivekanandan et al. (2009)

process known as "acclimation") or look for better exterior conditions. Fish species are extremely sensitive to temperature fluctuations because they have particular thermal preferences for different life phases, including spawning, growth, and reproduction. These preferences may be disrupted by rising water temperatures, requiring fish to adjust or suffer the consequences. The fish will naturally modify their internal metabolic processes through mechanisms including home phasic adaptation and home viscous adaptation in response to temperature fluctuations. The balance and proper operation of the organism depend on these cellular reactions. Behavioral thermoregulation is an additional mechanism for adapting to temperature fluctuations. According to Casterlin and Reynolds (1980), poikilothermic animals achieve thermoregulation through behavior by physically moving along a temperature gradient to reach the optimal temperature. Fish species are extremely sensitive to temperature fluctuations because they have particular thermal preferences for different life phases, including spawning, growth, and reproduction. These preferences may be disrupted by rising water temperatures, requiring fish to adjust or suffer the consequences.

1. **Impact on Growth:** In general, fish exhibit the ideal temperature for growth and survival (Brett, 1979; Gadomski and Caddell, 1991). Since many species' juveniles prefer to dwell in warmer climates than their adults, these vary with age and size (McCauley and Huggins, 1979; Pedersen and Jobling, 1989). Additionally, differing ideal temperatures may exist in early life stages, which could have an impact on temporal and spatial field distributions. It has also been reported that for a number of fish species, the combined effects of temperature and size on growth (Brett, 1979; Imsland et al., 2006). These temperature ends are, of course, not constant and are dependent on the acclimatization temperature to some extent. There can be noticeable variations in both the top and lower limits when a species can be adapted to very varied temperatures, such as a range of 20°C (Pörtner and Peck, 2010).
2. **Reproductive Difficulties:** Temperature variations can affect the timing and efficacy of fish reproduction. In several fish species, precise temperature cues signal the beginning of spawning. If

water temperatures vary or become less predictable and shift outside their preferred range, it may harm recruitment. This is because it may cause an overlap between spawning events and the availability of suitable food sources for fish fry.

3. **Changes in Distribution:** Coldwater species that were historically common in Indian rivers and lakes may see losses when temperatures rise. On the other hand, warm water species might prosper and widen their geographic ranges. This alteration may alter the make-up of fish communities, which will affect the biodiversity of freshwater ecosystems as a whole.

4. **Heat Stress Vulnerability:** High temperatures can also make fish more susceptible to infections and heat stress, which may result in higher mortality rates. Fish immune systems can be weakened by heat stress, leaving them more prone to parasites and diseases.

10.2.6 CHANGES IN PRECIPITATION PATTERNS

Climate change also influences precipitation patterns, leading to altered hydrological cycles in many regions of India. Increased rainfall intensity and frequency can result in sudden floods, destroying fish habitats and causing fish mortality. Conversely, prolonged droughts can reduce water levels, limiting fish access to suitable habitats and food sources. These changes manifest in two primary ways: increased rainfall intensity and frequency, resulting in floods, and prolonged droughts leading to reduced water levels.

Prolonged Droughts

1. **Reduced Water Levels:** Certain places may experience protracted droughts and a decline in precipitation as a result of climate change. Lower water levels in rivers, lakes, and reservoirs due to these droughts may make it harder for fish to access habitable habitats and food supplies. Fish could become increasingly concentrated in smaller amounts of water, which would increase competition for resources and increase their susceptibility to predators.

2. **Reduced Breeding Success:** Because of low water levels, fish may not be able to reach their usual breeding areas when there is a drought. Additionally, warmer water temperatures in shallower regions can impact fish eggs and fry survival.
3. **Water Quality Issues:** The growth of toxic algae species, which produce toxins detrimental to fish, can be sped up by higher temperatures. Furthermore, fish mortality brought on by decreased oxygen levels in warmer water bodies can harm fish populations as a whole Quality of Water Issues. The growth of toxic algae species, which produce toxins detrimental to fish, can be sped up by higher temperatures. Furthermore, fish mortality brought on by decreased oxygen levels in warmer water bodies can harm fish populations.
4. **Fish Toxic Effects:** Toxins released into the water by harmful algal blooms (HABs) can have a variety of harmful effects on fish. Fish exposed to or ingesting these poisons may endure stunted growth, poor reproductive outcomes, and even death. Fish tissues may also become contaminated with the poisons, rendering them unfit for human consumption.
5. **Disruption of Food Webs:** By poisoning or killing phytoplankton, the base of the food chain, HABs can disrupt aquatic food webs. Fish populations may be affected as a result of the disruption's cascade consequences on fish prey availability.

Oxygen Depletion

1. **Warmer Water and Oxygen Solubility:** The ability of warmer water to contain dissolved oxygen is lowered. Because of climate change, water bodies are less effective at keeping oxygen, which is necessary for fish survival, as temperatures rise.
2. **Fish Kills:** Fish are unable to collect enough oxygen from the water as a result of oxygen depletion, which can result in fish deaths. Particularly, susceptible to such occurrences are fish species with higher oxygen requirements or those already under stress from other environmental variables. Fish deaths can have enormous ecological and financial repercussions.
3. **Altered Fish Behavior:** Fish may alter their behavior in reaction to falling oxygen levels. They may seek regions with greater oxygen or shift to shallower waters, which can alter their usual distribution patterns and render them more vulnerable to predators.

Increased Erosion and Sedimentation

1. **Extreme Weather Events:** Storms and heavy rainfall may occur more frequently and with greater severity as a result of climate change. These occurrences may cause soil erosion in river catchment areas, which may result in the deposition of sediments into bodies of water.
2. **Smothering of Spawning Grounds:** These regions are frequently found in riverbeds or shallow areas with gravel or sand substrates. Fish larvae and fry have a harder time surviving when sediments cover these sites, making them unsuitable for fish reproduction.
3. **Reduced Water Clarity:** Excessive sediments in water bodies can cause the water to be less clear, making it harder for fish to find food and avoid predators. Their overall fitness and feeding efficiency may suffer as a result.

Rising Sea Levels

1. Saltwater Intrusion

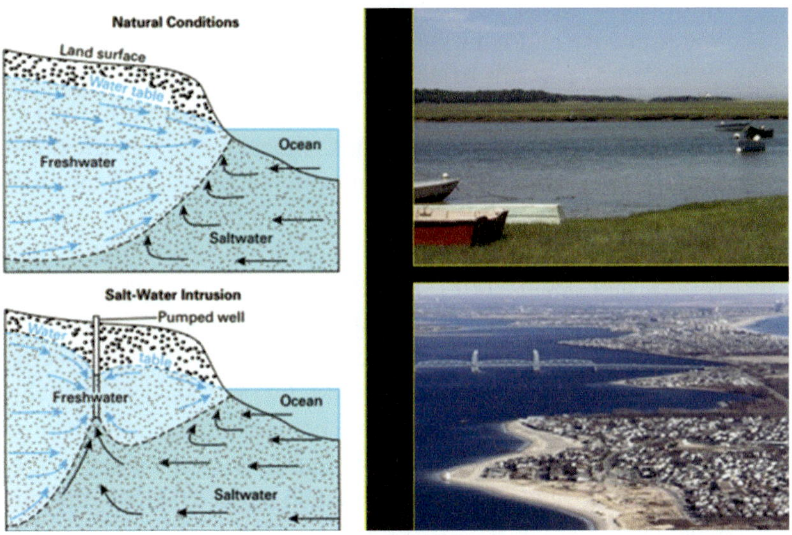

Source: USGS. https://www.usgs.gov/media/images/process-saltwater-intrusion

Effect on Coastal Waters: Increasing sea levels brought on by global warming may cause freshwater ecosystems in coastal locations to become contaminated with saltwater. Due to this incursion, previously freshwater habitats may become too salty for many fish species, resulting in the extinction of local freshwater species. The process of saltwater intrusion into an aquifer system is depicted in the above figure. The freshwater/saltwater interface is the line that separates fresh groundwater from saltwater. Saltwater cannot seep landward when fresh groundwater discharges to the coast. A phenomenon known as "saltwater intrusion" allows saltwater to flow landward if an excessive amount of freshwater is extracted from the aquifer system. A pumping well may allow saltwater to infiltrate and contaminate the water supply if it is located near a freshwater/saltwater interface that migrates landward.

2. **Loss of Estuarine Habitats:** Estuaries are crucial for finfish as nursery and feeding environments; however, in northern Europe, anthropogenic activities usually have a significant negative influence on them (Airoldi et al., 2008). Estuaries are extremely dynamic ecosystems that are under an extreme stress from both the natural world and humans (Attrill et al., 1999). Estuarine water quality and habitat extent can be significantly altered and degraded as a result of changes in sediment supply, hydrology, sea level rise, human activities in intertidal ecosystems, and land management in nearby locations. Since many significant fish species have specialized habitat preferences or confined migration patterns, ongoing habitat loss and degradation may have a detrimental effect on the population of fish within estuaries (Fodrie and Levin, 2008). Therefore, it has been suggested that, in comparison to historic levels, anthropogenic activities like those mentioned above have decreased the ability of estuarine ecosystems to support fish populations (Mclusky et al., 1992; Rochette et al., 2010). Estuaries and certain pertinent habitats, such as saltmarsh, are legally protected in the European Union and the United Kingdom by legislative policies (such as the Habitat's directive: Council Directive 92/43/EEC) or as Sites of Special Scientific Interest. These policies seek to lessen human disturbance and preserve or improve the ecological condition of designated habitats or species within site boundaries. Nevertheless, according to Vasconcelos et al. (2007), these site designations frequently do

not include dependent fish species or assemblages in management or monitoring plans.

Deforestation and Alterations in Land Use

1. **Watershed Modification:** Deforestation and changes in land use can modify watersheds' hydrological features. With less forest cover, there may be greater surface runoff, which would result in faster water flow into rivers and streams. This changed flow may interfere with water bodies' normal flow patterns, which may affect fish habitats.
2. **Loss of Riparian Vegetation:** Eliminating riparian vegetation along riverbanks may increase sediment runoff and limit water body shading. The stabilization of riverbanks, prevention of erosion, and provision of critical habitat for aquatic organisms are all made possible by riparian vegetation.
3. **Threats from Invasive Species:** Invasive fish species may spread more readily as a result of climate change. These invasive species have the potential to displace native fish, disturb ecosystems, and lessen fish diversity overall. Indigenous fish populations in India have already faced substantial difficulties from invasive species like the African catfish and Mozambique tilapia.

Competition with Native Species

1. **Competition for Resources:** Invasive fish species can compete with native fish for vital resources like food and breeding grounds. They might eat comparable prey or occupy comparable habitat niches, which might decrease the food supply for native species and degrade their environment.
2. **Predation:** Invasive fish species may feed on native fish, especially when native fish populations are already under stress due to habitat loss or other environmental disturbances. This predator pressure may result in even fewer native fish populations.

Management of sustainable fisheries

1. **Size Limits and Catch Quotas:** Setting size limits and catch quotas can help control fishing activity, stop overfishing, and promote the recovery of fish populations. By preventing fish from being caught

before they reach reproductive age, these techniques help to maintain sustainability.

2. **Seasonal Closures:** By establishing seasonal closures in vital spawning and breeding sites, you can safeguard fish throughout sensitive life phases, boosting recruitment and the general health of the fish population.
3. **Ecosystem-Based Management:** It is essential to modify fisheries management strategies to take the larger ecosystem into account. To make wise management decisions, this must be aware of how various species, habitats, and environmental elements interact.
4. **Riparian Restoration:** By restoring the riparian areas along rivers and lakes, it is possible to reduce sedimentation, stop erosion, and improve the habitat quality for fish and other aquatic life.
5. **Revegetation:** Planting native vegetation in and around bodies of water can help shorelines stay stable, offer shade, and encourage the establishment of important aquatic plants.
6. **Diversified Aquaculture:** Promoting the growth of fish species from a variety of families that are more suited to shifting climatic circumstances might help reduce reliance on a single species and increase resilience.
7. **Improved Pond Management:** Using climate-smart aquaculture techniques, such as optimizing stocking densities, controlling water quality, and utilizing renewable energy sources, can lessen the susceptibility of fish farms to problems caused by the climate.
8. **Selective Breeding:** Creating and employing fish strains that are more resistant to disease and temperature changes can increase the sustainability of aquaculture.

Early Warning Devices

1. **Weather Monitoring:** By setting up weather monitoring systems, fishermen can get timely information about altering weather patterns, enabling them to better plan their fishing operations and steer clear of harsh weather events.
2. **Flood and Drought Alerts:** The creation of flood and drought alarm systems can assist fishing communities in anticipating and responding to these catastrophes, thereby saving lives and minimizing financial damages.

Knowledge and Instructions

1. **Community Workshops:** Holding training sessions and workshops for fishing communities to raise knowledge of how climate change is affecting fisheries and to provide details on adaptation measures.
2. **Transfer of Knowledge:** By fostering communication between scientists and fishing communities, it is possible to help communities make appropriate climate adaptations.
3. **Promoting Sustainable Practices:** One way to increase the long-term resilience of fisheries is through educating fishermen about sustainable fishing methods and ethical resource management.

10.3 CONCLUSION

India's inland fisheries face severe threats from climate change, affecting ecological, social, and economic aspects. Collaborative efforts among policymakers, scientists, and local communities are crucial to safeguarding these ecosystems. Protecting India's valuable freshwater biodiversity and ensuring the well-being of millions depend on this collective endeavor. Climate change adaptation and mitigation are essential to secure the future of India's inland fisheries.

KEYWORDS

- **sustainability**
- **adaptation**
- **coastal**
- **climate change**
- **mitigation**
- **communities**

REFERENCES

Airoldi, L., Balata, D., & Beck, M. W. (2008). The gray zone: Relationships between habitat loss and marine diversity and their applications in conservation. *Journal of Experimental Marine Biology and Ecology*, *366*(1–2), 8–15.

Attrill, M. J., Bilton, D. T., Rowden, A. A., Rundle, S. D., & Thomas, R. M. (1999). The impact of encroachment and bankside development on the habitat complexity and supra-littoral invertebrate communities of the Thames estuary foreshore. *Aquatic Conservation: Marine and Freshwater Ecosystems*, *9*, 237–247.

Bellard, C., Bertelsmeier, C., Leadley, P., Thuiller, W., & Courchamp, F. (2012). Impacts of climate change on the future of biodiversity. *Ecology Letters*, *15*(4), 365–377.

Brett, J. R. (1979). Environmental factors and growth. In: Hoar, W. S., Randall, D. J., & Brett, J. R., (Eds.), *Fish Physiology*, Academic Press: London, Vol. 8, pp. 599–675.

Casterlin, M. E., & Reynolds, W. W. (1980). Thermoregulatory behavior of a tropical marine fish: Canthigaster jactator (Jenkins). *Hydrobio*, *70*, 269–270.

CMFRI. (2014). *Annual Report*. Central Marine Fisheries Research Institute: Kochi, India.

FAO. (2020). *The State of World Fisheries and Aquaculture 2020: Sustainability in Action*. Food and Agriculture Organization of the United Nations.

FAO. (2022). The state of world fisheries and aquaculture 2022. In: *Towards Blue Transformation*. FAO: Rome. https://doi.org/10.4060/cc0461en.

Fodrie, F. J., & Levin, L. A. (2008). Linking juvenile habitat utilization to population dynamics of California halibut. *Limnol Oceanogr*, *53*(2), 799–812

Gadomski, D. M., & Caddell, S. M. (1991). Effects of temperature on early-life-history stages of California halibut Paralichthys californicus. *Fishery Bulletin*, *89*, 567–576.

Imsland, A. K., Foss, A., Sparboe, L. O., & Sigurðsson, S. (2006). The effect of temperature and fish size on growth and food efficiency ratio of juvenile spotted wolffish. *Journal of Fish Biology*, *68*, 1107–1122.

McCauley, R. W., & Huggins, N. W. (1979). Ontogenetic and non-thermal seasonal effects on thermal preferenda of fish. *American Zoologist*, *19*(1), 267–271.

McLusky, D. S., Bryant, D. M., & Elliott, M. (1992). The impact of land-claim on macrobenthos, fish and shorebirds on the forth estuary, eastern Scotland. *Aquatic Conservation: Marine and Freshwater Ecosystems*, *2*(3), 211–222.

Parmesan, C. (2006). Ecological and evolutionary responses to recent climate change. *Annual Review of Ecology, Evolution, and Systematics*, *37*, 637–669.

Pedersen, T., & Jobling, M. (1989). Growth rates of large, sexually mature cod Gadus morhua, in relation to condition and temperature during an annual cycle. *Aquaculture*, *81*(2), 161–168.

Pörtner, H. O., & Peck, M. A. (2010). Climate change effects on fishes and fisheries: Towards a cause-and-effect understanding. *Journal of Fish Biology*, *77*(8), 1745–1779.

Prasanna Kumar, S., Roshin, R. P., Narvekar, J., Dinesh Kumar, P. K., & Vivekanandan, E. (2010). Signatures of global warming and regional climate shift in the Arabian Sea. In: Joseph, A., Nandan, S.B., Augustine, A. (Eds.), *Climate Change and Aquatic Ecosystems*, Cochin University of Science and Technology, Cochin, India, pp. 55–62

Rochette, S., Rivot, E., Morin, J., Mackinson, S., Riou, P., & Le Pape, O. (2010). Effect of nursery habitat degradation on flatfish population: Application to Solea in the Eastern Channel (Western Europe). *Journal of Sea Research*, *64*(1–2), 34–44.

Thompson, J., Charpentier, A., Bouguet, G., Charmasson, F., Roset, S., Buatois, B., & Gouyon, P. H. (2013). Evolution of a genetic polymorphism with climate change in a Mediterranean landscape. *Proceedings of the National Academy of Sciences*, *110*(8), 2893–2897.

Vasconcelos, R. P., Reis-Santos, P., Tanner, S., Fonseca, V., Latkoczy, C., Günther, D., & Cabral, H. (2007). Discriminating estuarine nurseries for five fish species through otolith elemental fingerprints. *Marine Ecology Progress Series*, *350*, 117–126.

Vivekanandan, E. (2011). *Climate Change and Indian Marine Fisheries*. CMFRl Special Publication: India.

Vivekanandan, E. (2013). Climate change: Challenging the sustainability of marine fisheries and ecosystems. *Journal Aquatic Biology and Fisheries*, *1*, 58–71.

Vivekanandan, E., Ali, M. H., Jasper, B. & Rajagopalan, M. (2009). Vulnerability of corals to warming of the Indian seas: A projection for the 21st century. *Current Science*, *97*, 1654–1658.

CHAPTER 11

Impacts of Microplastic on Aquatic Environment

PINAK BAMANIYA[1*], D. T. VAGHELA[1], ISHITA BAMBHANIYA[2], MAYUR BHADARKA[1], NITIN SUYANI[3], H. L. PARMAR[3], B. SUREANDIRAN[3], and RAJKUMAR DABHI[4]

[1]*Department of Aquatic Environment Management, College of Fisheries Science, Kamdhenu University, Veraval, Gujarat, India*

[2]*Department of Aquaculture, College of Fisheries Science, Kamdhenu University, Gujarat, India*

[3]*Department of Fisheries Resource Management, College of Fisheries Science, Kamdhenu University, Veraval, Gujarat, India*

[4]*Department of Aquaculture, College of Fisheries Science, Kamdhenu University, Navsari*

*Corresponding author

ABSTRACT

Microplastic (MP) pollution in aquatic environments poses significant ecological and human health concerns. Sampling techniques for MPs analysis vary, influencing abundance estimation. Freshwater bodies, including rivers and lakes, act as reservoirs for MPs from stormwater runoff, atmospheric deposition, and wastewater discharge. Urban areas contribute significantly to pollution through stormwater systems and untreated wastewater discharge. Bioaccumulation and biomagnification of MPs occur within aquatic food chains, impacting organisms and human health. MPs, due to their hydrophobic nature, absorb and transport hazardous substances, leading to genetic toxicity and ecosystem disruption.

Advances in Freshwater Aquaculture: Technology for Sustainable Development. Tarang Kumar Shah, Madhu Sharma, Mohd. Ashraf Rather, & Avdhesh Kumar (Eds.)

DOI: 10.1201/9781779640857-11

Estuaries play a crucial role, posing risks of bioaccumulation. Human exposure to MPs occurs through seafood consumption, a significant ingestion route. Aquaculture lacks comprehensive research on MP presence in farmed versus wild seafood. Monitoring methods, including microscopy and spectroscopy, are vital for understanding MP characteristics. In India, marine environments, sediments, beaches, and organisms face MP contamination. Research highlights correlations between human activities, tourism, and MP abundance on beaches. Fish, essential for coastal populations, exhibit MP ingestion, potentially causing adverse health effects. Comprehensive studies are essential for effective mitigation, emphasizing the need for continued research and global cooperation.

11.1 INTRODUCTION

Plastics have revolutionized numerous aspects of our lives, but the disposal of single-use plastics has become a major environmental issue. Approximately 10% of global plastic waste ends up in the oceans, eventually breaking down into microplastics (MPs). These tiny particles endure for extensive periods and pose a significant threat to marine life. MPs do not just indirectly enter oceans; they also directly flow through wastewater, underscoring the role of human activities in this problem. Addressing these entry points is crucial to safeguard our oceans.

Marine creatures ingest these MPs, impacting their growth and passing these pollutants up the food chain to larger animals. Consequently, seafood becomes a significant source of human exposure to MPs. Understanding and addressing this issue is pivotal for devising effective policies to protect marine life and those reliant on marine resources. Activities like fishing, shipping, and tourism also contribute to MP pollution in aquatic environments. Aquaculture systems, used for fish farming, often employ materials prone to degradation, leading to the generation of MPs during cleaning processes. Managing this pollution in aquatic systems presents a complex challenge.

MPs and pollutants negatively affect the reproductive capacities of aquatic organisms, resulting in dwindling population numbers. This decline significantly impacts the quantity and quality of harvested seafood, emphasizing the critical link between environmental health, aquatic life cycles, and productivity. Safeguarding the reproductive abilities of these organisms is vital for maintaining the sustainability of aquatic ecosystems and the industries they support (Table 11.1).

TABLE 11.1 MPs Polymer Present in Marine Environment

Polymers	Sources
Polyethylene (PE)	Bottle caps, cosmetics, disposal caps, fishing gear, household items, packaging films, plastic bags and storage containers, etc.
Polypropylene (PP)	Bottle caps, straws, fishing nets, fishing ropes, household appliances, etc.
Polystyrene (PS)	Disposable cutleries, plastic cups, plastic utensils, fishing gears, packaging films, thermocol boxes, etc.
Polyethylene tetrapthalate (PET)	Plastic bottles, packaging plastics, plastic containers, etc.
Polyvinyl chloride (PVC)	Construction materials, fishing materials, paints, pipes, plastic coatings, vinyl coating, etc.
Polyester	Fibers in clothings, textiles, and FRP boats

11.2 CLASSIFICATION OF MP

According to the National Oceanic and Atmospheric Administration, MPs encompass any plastic particles smaller than 5 mm in size (Arthur et al., 2009). Researchers typically categorize MPs into primary and secondary classifications (Andrady, 2011). Primary MPs are micronized and serve various purposes, such as in the cosmetic industry, clothing, fishing nets, among others. In contrast, secondary MPs result from the mechanical and oxidative breakdown of larger plastic materials like bottles and bags (Hidalgo-Ruz et al., 2012). Studies indicate that secondary MPs, originating from coastal beaches, are the most prevalent group in the environment (Andrady, 2011). These findings highlight coastal beaches as significant contributors of secondary MPs in marine environments.

The classification of MPs based on their shapes, sizes, and colors varies across different studies. For instance, researchers have categorized MPs into distinct shapes such as fragments, fibers, foams, beads, and pellets (Lusher et al., 2017). However, other studies like Karami (2017) and Free et al. (2014) propose different classifications, including types like fibers, films, fragments, and foams, or pellets, film, fragment, fiber, and foam, respectively (Karami et al., 2017; Free et al., 2014).

Similarly, the categorization of MPs according to size also differs. Daniel et al. (2020) divided MPs into six size groups ranging from 100 to 5000 μm, while Gurjar et al. (2021) classified them into five categories spanning from <100 μm to >1000 μm (Daniel et al., 2020; Gurjar et al., 2021).

Researchers have also attempted to classify MPs based on color composition, resulting in various categorizations. Young and Elliot (2016) identified seven color categories: blue, green, white/transparent, black/grey/brown, yellow/orange, red/pink, and violet/purple. In another classification by Li et al. (2015), MPs were grouped into five colors: red, black, blue, transparent, and brown (Young and Elliott, 2016; Li et al., 2015).

11.3 SOURCES OF MP

11.3.1 CLEANING PRODUCTS

The cleansing of surfaces often involves the utilization of primary MPs as abrasives. Plastics such as polyester, polycarbonate, and polyamide (PA) find application as blast media for this purpose. The primary objective of employing these blasting abrasives is multifaceted, encompassing tasks such as the thorough cleansing of rims, the efficient removal of paint coatings, and the meticulous cleaning of ships.

11.3.2 SYNTHETIC TEXTILES

The laundry processes conducted in both households and industrial settings contribute significantly to the mass production of primary MPs. This phenomenon arises from the scraping and dispersion of synthetic fibers during the laundering of textiles. These fibers, primarily composed of materials such as polyester and nylon, are released into the washing water as microscopic particles.

11.3.3 PERSONAL CARE PRODUCTS

Plastic microbeads serve a dual purpose as versatile additives in numerous cosmetics and personal care products. Functioning both as sorbents and exfoliators, these tiny beads are intentionally incorporated into various formulations to enhance product performance and consumer experience.

11.3.4 PAINTS AND WOOD PRESERVATIVES

Primary MPs play a crucial role when incorporated into paints and preservatives, serving various functional purposes that enhance the overall performance and longevity of treated surfaces, particularly wood.

11.3.5 SYNTHETIC WAXES AND OIL-GAS INDUSTRY

Synthetic waxes undergo scrutiny as primary MPs due to their diverse applications in various industries. These versatile substances are utilized for their unique properties in different processes, ranging from dyeing to food coating, acting as lubricants, contributing to plastic processing, and playing a crucial role in textile processing.

11.3.6 PACKAGING MATERIALS

Packaging materials are indeed a significant contributor to the presence of MPs, constituting a substantial portion of plastic waste globally. This preference for packaging materials as a source of MPs is underscored by their widespread use and disposal practices.

11.3.7 DOMESTIC ITEMS

Domestic items, particularly single-use plastics, significantly contribute to marine pollution. Items such as cups, plastic cutlery, bottles, and straws are found abundantly in oceans and pose a significant threat to marine ecosystems.

11.4 FACTORS AFFECTING ABUNDANCE OF MPS

MPs found in both freshwater and marine settings can stem from diverse sources, whether land-based or marine-related. However, studies indicate that approximately 80% of marine debris originates from activities on land, with the remaining 20% attributed to marine-based sources (Jang et al., 2020). Factors such as population density and proximity to urban areas are believed to significantly impact the presence of MPs in these environments.

Increased levels of MPs have been linked to occurrences of rainfall. Investigations conducted in Swiss lake surface waters and Lake Ontario sediments revealed a rise in MP concentrations following rainfall events (Corcoran et al., 2015). One possible rationale behind this phenomenon is that runoff from land after rainfall, where plastic production and degradation rates are higher, could transport these MPs into aquatic environments (Dris et al., 2016). Consequently, heavier rainfall might intensify the erosive impact of land runoff, leading to the transportation of more plastic debris. Moreover, hydrodynamics plays a role in redistributing MPs within water bodies. Sediments tend to accumulate higher amounts of MPs compared to surface water (Zhang et al., 2018). Therefore, increased hydrodynamics during rainfall could cause the resuspension of microplastic debris from sediment, contributing to their presence in surface water. Additionally, prior studies have noted the presence of MPs in the atmosphere (Klein and Fischer, 2019). A hypothesis suggests that these MPs might adhere to raindrops and directly enter lake water during rainfall events.

Spillage from domestic and industrial sewage is considered a significant potential source of MPs in both marine and freshwater systems. However, the extent of contribution from these sources remains a topic of debate. Carr et al. (2016), for instance, found minimal detection of MPs in the effluents of a tertiary wastewater treatment plant (WWTP) (Carr et al., 2016). Similarly, the discharge from a secondary WWTP exhibited low MP abundance (one plastic particle per 1.14 liters of effluent). In contrast, Murphy et al. (2017) reported that although a secondary WWTP in Scotland achieved a removal rate of 98.41%, it still discharged 6.5 × 107 MPs daily (Murphy et al., 2017). This discrepancy in findings suggests that different treatment methods employed in each plant could result in varying removal rates, as well as quantitative differences in the size and type of MPs present. Research by Mintenig et al. (2019) demonstrated that MPs smaller than 500 micrometers in diameter were more prevalent (ranging from 10 to 9000 particles/m^3) compared to larger plastic particles (ranging from 0 to 50 particles/m^3) across 12 WWTPs in Germany (Mintenig et al., 2019). Additionally, PE emerged as the most abundant polymer in both size categories, followed by polyester fiber with concentrations ranging from 90 to 1000 particles/m^3.

The choice of sampling techniques when collecting environmental samples for MPs analysis, particularly in aquatic environments, significantly influences the estimation of their abundance (Alimi et al., 2020). However, different sampling and analytical methods can lead to either an underestimation or an overestimation of MPs abundance. For example, manta trawl-based sampling methods enable the collection of large water volumes in a relatively short time. Nevertheless, the mesh size of the net (typically 300–500 μm) may result in an underestimation of MPs concentration in the environment (Lindeque et al., 2020). Additionally, the materials used in the sampling device itself—such as frames and connectors composed of various plastic polymers like PA, PE, PP, or PVC—pose the risk of sample contamination.

11.5 DISTRIBUTION AND TRANSPORT OF MPS

11.5.1 MP IN THE FLUVIAL ENVIRONMENT

Plastic pollution poses a significant threat to both inland and marine environments. In inland waters, urban lakes, and riverbanks, plastic pollution is prevalent, with lakes acting as sinks for MP. In marine environments, areas with ocean current convergence, beaches, and seafloors are potential deposition sites for MP. The movement of MP is influenced by water currents, making it challenging to pinpoint sources precisely. Deep marine sediments are considered the ultimate repository for MP pollution, with higher concentrations compared to terrestrial soil. Interestingly, plastics can move in both directions—from land to water and water to land. Ocean–atmosphere exchange contributes to the presence of MP in soils, with winds transporting MP from land to sea and wave action carrying them back to land. MP can also accumulate in various environments, including vegetation in ecosystems like mangroves and seagrass beds.

MPs are pervasive in global freshwater environments, stemming from various origins like atmospheric deposition, groundwater seepage, stormwater runoff, and industrial or municipal wastewater discharge. Concentrations differ across locations, with urban streams and glaciers often exhibiting elevated levels. Effectively managing and treating industrial and municipal wastewater becomes pivotal in curbing the inflow of MPs into surface water bodies.

These minute plastic particles permeate freshwater bodies, including rivers, lakes, and streams, and their dispersion is influenced by factors like wind, currents, flow rates, and temperature. In freshwater systems, the flow rate governs the spatial distribution of MPs—slower flows lead to sediment buildup, while higher flows aid transport and potential resuspension.

Lakes and low-flow urban streams act as reservoirs for accumulating MPs, while rivers function as conduits transporting MPs toward the ocean. Urban water bodies, in particular, amass secondary MPs. Freshwater systems, especially rivers, serve as gathering points for MPs sourced from diverse origins like wastewater, stormwater, runoff, and atmospheric deposition.

MP composition in rivers displays variability, notably in urban areas where higher fiber content is prevalent, often originating from textile products. The shape of fibers, conducive to floating and aggregation during transportation, significantly affects their destiny in surface waters. This underscores the substantial role of freshwater systems in the dissemination and distribution of MPs within the environment.

Urban areas in the western United States witness stormwater collection systems that gather runoff containing sediment, silt, and debris from precipitation. These systems, often untreated, discharge directly into water bodies, contributing to the release of contaminants, including MPs, into lakes, streams, and bays. MPs can infiltrate the land surface through various means such as littering, aerial deposition, intentional applications (like biosolids and agricultural amendments), and product degradation like tire wear particles. Once on land, these plastics can be carried by stormwater runoff, eventually flowing into collection systems, lakes, streams, or low-lying areas.

Laboratory experiments conducted by Koutnik et al. (2022) have suggested that freeze-thaw cycles might potentially drive denser MPs (e.g., PET and PVC) deeper into the subsurface beneath stormwater treatment systems, posing a heightened risk of groundwater pollution (Koutnik et al., 2022).

Bays and estuaries, serving as transitional zones between aquatic and terrestrial habitats, are pivotal for the fate and movement of MPs. They serve as repositories for larger plastic debris and sources of MPs, particularly influenced by the surrounding human population. Mangrove swamps within estuaries, crucial for coastal protection and filtration, significantly accumulate MPs compared to other environments, influencing soil properties.

These estuarine ecosystems, acting as MP repositories, pose a risk of bioaccumulation in the food chain. Crustaceans, mussels, and deposit feeders face an increased risk of ingesting MPs, which can alter nutrient cycling and potentially contribute to harmful algal blooms. The interconnection of these ecosystems with human livelihoods underscores the necessity for comprehensive studies on the impact of MP in these crucial coastal areas.

MPs originating from industrial and domestic products, like synthetic clothing, agricultural items, and cleansers, enter municipal wastewater systems. WWTPs play a vital role in removing MPs, but their efficiency varies from 50% to 99.9%. Despite treatment, biosolids concentrate MPs, raising concerns when these materials are recycled through land application or composting.

Significant global untreated wastewater discharge (50–80%) lacks standardized processes specifically targeting MP removal. Recent studies in Mauritius found that over 90% of MPs in wastewater effluents were smaller than 0.5 mm, raising concerns about their potential entry into the food chain. Responding to this, Mauritius implemented a ban on plastic bags in 2015. Another study in Canada revealed the discharge of 141 million MP particles daily from a Class 4 WWTP into the South Saskatchewan River, indicating potential gaps in MP removal processes (Petersen and Hubbart, 2021; Ragoobur et al., 2021; Prajapati et al., 2021).

11.6 BIOACCUMULATION AND BIOMAGNIFICATION

Bioaccumulation and biomagnification are fundamental concepts in ecological risk assessments that traditionally pertain to the transport of dissolved chemical contaminants within food chains (Mackay and Boethling, 2000; Alexander, 1999). Although originally associated with dissolved chemical contamination, these terms have been adopted in the context of MP literature (Dawson et al., 2018; Akhbarizadeh et al., 2019; Rochman et al., 2019). In this context, bioaccumulation, or body burden, is defined as the overall uptake of a contaminant—such as MPs or additives—from the environment through various routes (e.g., contact, ingestion, respiration) and from any source (e.g., water, sediment, prey)

(Gerber, 2009; Simpson et al., 2016). Simply put, bioaccumulation occurs when an organism absorbs a contaminant at a rate higher than its ability to eliminate it (Wang et al., 2016). This process can lead to the subsequent transfer of the contaminant up the food chain, resulting in its biomagnification at higher trophic levels (Kelly et al., 2007). Biomagnification along a food web refers to the increase in the concentration of a contaminant—like MPs or additives—in an organism compared to its prey (Gerber, 2009; Simpson et al., 2016). It is important to note that this definition assumes that contamination in higher trophic levels is a direct consequence of consuming prey from lower trophic levels, that is, trophic transfer is taking place.

11.6.1 MP TRANSFER THROUGH FOOD CHAINS

MPs possess the capability to absorb and transport various substances, including chemicals and pathogens, within the human food chain (Schmaltz et al., 2020). Beyond their direct ingestion and inherent toxicity in organisms MPs, owing to their hydrophobic and lipophilic nature along with a high surface-to-volume ratio, can adsorb toxic pollutants like bisphenol A (BPA), monomers, refractory chemicals, heavy metals, antibiotics, and oligomers (Guo et al., 2020; Thushari and Senevirathna, 2020). This enables the transfer of these pollutants to the tissues of organisms across the food chain levels, amplifying risks for both organisms and humans (Thushari and Senevirathna, 2020).

MPs can carry hazardous elements like DDT and hexachlorobenzene. The ingestion of MPs containing chemicals such as mercury and polycyclic aromatic hydrocarbons, originating from composite plastic chemicals, leads to genetic toxicity in organisms, posing a threat to the flora and fauna in their habitats (Zhang et al., 2020). Persistent organic pollutants like organochlorine pesticides, dioxins, and polychlorinated biphenyls that resist biodegradation are known to adsorb onto the surface of MPs.

These particles have a propensity to accumulate in organisms through the food chain, negatively impacting reproduction, chromosome mutation, and growth due to substances like phthalates and BPA. Furthermore, antibiotics such as sulfamethoxazole and trimethoprim, adhering to MPs like PE, PP, PA, PET, PS, and PVC, can alter bacterial populations, potentially endangering both organisms and human health (Thushari and Senevirathna, 2020). The potential release of additives used in plastic production into water via MPs may also adversely affect the endocrine function of marine organisms (Bai et al., 2021).

11.7 ECOLOGICAL IMPACTS

11.7.1 OCCURRENCE AND EFFECTS OF MICRO- AND NANOPLASTICS (MNPS) IN AQUATIC HABITATS

MNP pollution is prevalent in aquatic (both marine and freshwater), terrestrial, and atmospheric environments. These environments are linked through various source-pathway-sink networks, affecting the movement and accumulation of MNPs (Zhang et al., 2020; Bergmann et al., 2019). Despite this interconnectedness, research over the past decade has largely focused on aquatic environments rather than terrestrial or atmospheric ones. MNPs in water primarily consist of plastics like PE, PP, PS, and PVC, which interact with organic matter, inorganic elements, and microorganisms (Mammo et al., 2020). Studies have shown that MNPs can absorb toxic chemicals, effectively acting as carriers for these substances across different environments. Consequently, MNPs can be ingested by marine organisms, gradually infiltrating the marine food web and posing significant risks to both marine and terrestrial life (Al-Thawadi, 2020).

11.7.2 EFFECTS ON AQUATIC LIFE

MPs pose a global concern due to their harmful effects on both fish and humans. As fish is a crucial source of protein for human consumption, contamination of fish by MPs becomes a significant hazard that demands particular attention. Fish exposed to MPs, either by themselves or alongside other pollutants, can experience a range of health problems. MPs can inflict tissue damage, induce oxidative stress, alter immune-related gene expression, and affect antioxidant status in fish. Exposure to MPs may lead to neurotoxicity, hindered growth, and behavioral abnormalities in fish. However, the implications of MPs on human health remain poorly understood. Given the widespread presence of MPs in the environment, exposure can occur through consumption, inhalation, and skin contact.

Human exposure to MPs can lead to oxidative stress, cytotoxicity, neurotoxicity, immune system disruption, and the migration of MPs to various tissues.

11.7.3 BIOAVAILABILITY OF MPS TO FISH

MPs, prevalent in various aquatic environments, share sizes akin to sediments and certain planktonic species, making them accessible to a wide array of aquatic creatures, notably fish (Pazos et al., 2017; Kumkar et al., 2021; Sequeira et al., 2020). The availability of MPs to fish is influenced by several factors. Filter-feeders and bottom-dwelling fish are believed to be more susceptible to MPs ingestion due to their indiscriminate feeding behavior compared to predator fish (Wesch et al., 2016; Lusher et al., 2020). Mizraji et al. (2017) explored the link between tidal fish-eating patterns and MPs ingestion, finding that omnivorous fish consumed more MPs than herbivorous and carnivorous fish (Mizraji et al., 2017).

MPs resembling natural prey are more likely to be ingested by visually oriented fish species. *Pomatoschistus microps* showed a selective ingestion of white MPs over black and red particles (de Sá et al., 2015). Additionally, the shape and size of MPs can enhance their likelihood of being consumed by fish (Auta et al., 2017). Boerger et al. (2010) found that the Myctophidae family of fish in the North Pacific Central Gyre typically ingested MPs ranging from 1 to 2.79 mm in size, which is similar to the size of plankton, their main food source (Boerger et al., 2010).

11.7.4 IMPACTS OF MPS, ON FISH

The fish used in MPs' exposure experiments were sourced from various environments, primarily from marine habitats. Accumulation of MPs in the gastrointestinal tract of fish following ingestion can lead to blockages along the digestive system, reducing feeding behavior due to decreased appetite (Lusher et al., 2013; Wright et al., 2013). MP intake might also prompt structural and functional alterations in the digestive tracts of fish, resulting in dietary and developmental challenges (Huang et al., 2022; Jabeen et al., 2018; Borrelle et al., 2017; Peda et al., 2016). Studies, primarily conducted on Daniorerio, commonly highlight oxidative stress, reduced mobility, disruptions in gene expression, and damage to reproductive organs as predominant effects of MPs exposure in this species (Mu et al., 2021; Zhao et al., 2021; Zhang et al., 2022). Figure 11.1 indicates the possible dangerous effects of MPs injection into fish.

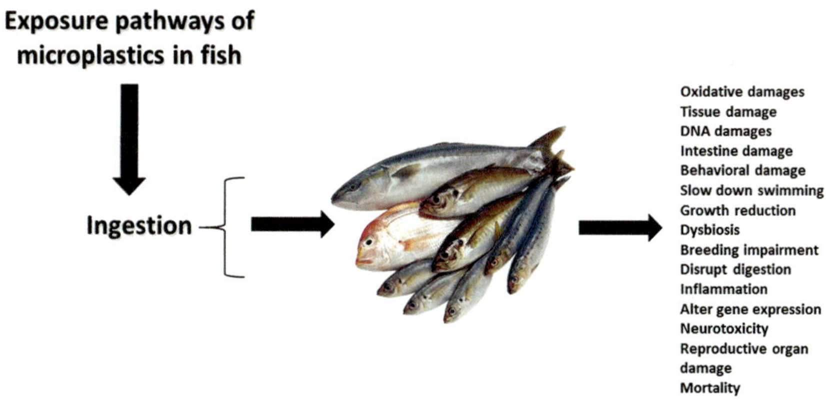

FIGURE 11.1 Exposure pathways of MPs in fishes

11.8 HUMAN HEALTH CONCERNS

Seafood consumption stands as a route for human exposure to MPs. As of 2015, global seafood intake accounted for 6.7% of total protein consumed and around 17% of animal protein intake (FAO, 2016). The average global per capita seafood consumption exceeds 20 kg per year, while in the United States, it stands at 7 kg annually (Fisheries of the United States, 2015). The 2016 global seafood trade amounted to $132.6 billion, with more than 90% of US seafood sourced from regions with significant waste leakage and widespread plastic pollution in open waters (Lusher et al., 2017). About half of the seafood comes from aquaculture (e.g., farmed fish) and the other half from wild catches.

Aquaculture enables environmental control by raising animals in controlled settings like ponds, tanks, or specific water bodies. Moreover, animals in aquaculture generally have shorter lifespans compared to those in the wild, potentially offering fewer opportunities and less time for MP exposure and absorption. However, due to limited research, there remains uncertainty regarding the differences in MP presence between farmed and wild fish and shellfish.

MPs, owing to their minute size, are susceptible to ingestion by a diverse range of marine creatures. This ingestion can occur directly or indirectly through trophic transfer, where MPs move up the food web. Evidence of MP ingestion has been recorded in planktonic organisms and larvae, which occupy the lower tiers of the food chain (Prajapati et al., 2021; Mackay and Boethling, 2000; Alexander, 1999; Dawson et al., 2018). Additionally, both small and large invertebrates, as well as fish, have been documented to ingest MPs (Lusher et al., 2017; MCole et al., 2013; Lee et al., 2013; Steer et al., 2017; Thompson et al., 2004). Observations have highlighted the trophic transfer of MPs in predatory Crucian carps (Duis and Coors, 2016).

MPs are present in various species meant for human consumption, including invertebrates, crustaceans, and fish (Farrell and Nelson, 2013; Murray and Cowie, 2011). These plastic particles are often concentrated in the digestive tracts of organisms, making bivalves and small fish, which are consumed whole, more likely to introduce MPs into the human diet (Mattsson et al., 2015). For instance, Van Cauwenberghe and Janssen found that farmed mussels contained significantly higher concentrations of MPs (178 microfibers) compared to wild-caught mussels (126 microfibers) (Farrell and Nelson, 2013). Additionally, Rochman et al. discovered MPs (> 500 μm) in commercially sold wild-caught fish from markets in Makassar, Indonesia (28% of fish processed contained MPs), and California, USA (25% of commercial fish processed contained MPs) (Murray and Cowie, 2011). Karami et al. investigated the potential presence of MPs in dried fish tissue, including excised organs (viscera and gills) and eviscerated flesh (whole fish excluding the viscera and gills) (Karami et al., 2017; Mattsson et al., 2015). They found that in 4 out of 30 commonly consumed dried fish species, 36 out of 61 isolated foreign particles were identified as plastic polymers (Van Cauwenberghe and Janssen, 2014). Yifeng et al. demonstrated that MP particles can translocate from the digestive tracts to the gills and liver in both young and adult zebra fish (Daniorerio), a common prey fish (Lu et al., 2016; Rochman et al., 2015). This translocation has also been documented in European seabass and the common goby *(Pomatoschistus microps*) (Lu et al., 2016). These studies collectively show the presence of MPs in some seafood, highlighting the potential for a widespread issue due to environmental ubiquity and the movement of particles to parts of the animal typically consumed by humans.

11.9 MONITORING AND DETECTION METHODS OF MP

Various techniques have been employed for the identification of MPs and nanoplastics (NPs) in aquatic ecosystems, terrestrial environments, food, as well as animal and human tissues.

11.9.1 MICROSCOPY (LIGHT)

The visual detection of MPs and NPs traditionally relies on light microscopy, assessing their size, shape, and color. The characterization of these particles through light microscopy primarily involves examining morphological and physical properties, tailored to the research objectives. Size categories typically range among different polymer species: <100 µm (62–100 µm), 100–500 µm, or > 1000 µm. However, this technique does not determine polymer identity. Still, various image processing programs like Histolab and Dinocapture can aid in quantifying the abundance of MPs and NPs once putative plastic fragments are observed (Wang and Wang, 2018; Desforges et al., 2014). Though effective, microscopy has limitations—it can be subjective, labor-intensive, and reliant on the observer. Automation and signal processing, facilitated by software, aim to mitigate these limitations. Dyes like eosin B, rose bengal, and Hostasol Yellow 3G are frequently used to improve visual observation. The success of stereomicroscopy depends on the size and shape of plastic particles; for instance, fibrous-shaped MPs and NPs are easier to observe, while debris <500 µm poses challenges in characterization. This technique lacks the ability to distinguish natural biopolymers (e.g., chitin) from human-made synthetic particles. Attempts have been made to apply physical stress using spatulas to confirm plastic identities, primarily in citizen science applications, but this method is not recommended for other purposes (Dris et al., 2017; Cai et al., 2017).

11.9.2 SCANNING ELECTRON MICROSCOPY (SEM)

SEM (involves utilizing a high-intensity electron beam to scan the surface of target materials. This technique, complemented by energy dispersive X-ray spectroscopy, is effective not only in identifying MPs and NPs but also in detecting any attached organic pollutants on their surfaces. Samples are visualized when their surfaces conduct electricity, often achieved by coating them with conductive materials like palladium, platinum, tungsten, or gold. SEM has proven successful in differentiating MPs ranging from 1 μm to 1 mm in size (Rocha-Santos and Duarte, 2015; Crawford and Quinn, 2016).

However, SEM's accuracy in identifying the polymer types of MPs and NPs might be compromised by chemical additives and coatings applied during production. This technique involves high capital and maintenance costs, demands skilled operators, and requires extensive sample preparation. Additionally, SEM analysis requires specific space conditions to avoid interference from electromagnetic and vibrational noise generated by nearby instruments and activities (Crawford and Quinn, 2016).

11.9.3 FOURIER TRANSFORM INFRARED SPECTROSCOPY

Fourier transform infrared spectroscopy (FTIR) serves as a key analytical method used to recognize synthetic polymers by examining their distinct molecular vibration spectra. It stands as a nondestructive tool for characterizing both NPs and MPs taken from environmental samples. Fu et al. discussed four primary FTIR analysis techniques—transmission, diffuse reflectance, true specular-adsorption, and attenuated total reflection—which are employed in this field (Cauwenberghe et al., 2013).

FTIR analysis leverages discriminatory features within spectra and reference libraries available online to conveniently identify common polymer types, providing detailed data on particle size. Commonly detected polymer types in various environmental samples like soil sediment and wastewater include PA, PE, PET, polymethyl methacrylate, PP, PS, and PVC (Fries et al., 2013; Fu et al., 2020; Browne et al., 2011; Claessens et al., 2011; Harrison et al., 2012). The sizes of these polymers in samples range from <1 to 1000 mm (Nor and Obbard, 2014). FTIR boasts a spatial resolution of 10–20 μm, making it one of the most favored spectroscopy methods. Micro-FTIR specifically analyzes MPs and NPs from dust and atmospheric samples, while attenuated total reflectance FTIR—a derivative methodology—is gaining attention for nondestructively measuring MPs and NPs with diverse shapes and sizes.

FTIR's nondestructive nature allows its coupling with other quantification techniques to comprehensively characterize polymers. However, its widespread use is restricted by high capital costs, limited accessibility to instruments, and the need for skilled operators to interpret spectra. Additionally, the chemical digestion process in plastic analysis pipelines may influence FTIR spectra's vibrational bands, causing spectral shifts. Moisture traces in samples can also interfere with the analysis, posing additional challenges.

11.9.4 RAMAN SPECTROSCOPY

Raman spectroscopy, based on the inelastic scattering of photons typically from a laser, is a nondestructive technique providing structural insights into larger polymers. It is notably effective in identifying small MPs (<20 µm) (Vianello et al., 2013). When combined with microscopy, it is termed micro(µ)-Raman and is valuable in detecting both MPs and nanoparticles (Browne et al., 2010). Like FTIR, Raman spectroscopy reveals structural details, showcasing polymer identity and various additives without damaging the sample. It offers benefits such as minimal sample requirements and the ability to assess the structure of the screened polymers (Smith and Dent, 2019; Lee and O'Donnell, 2019; Weber and Merlin, 2000).

Raman's resolution is limited to larger nanoparticles, yet recent advancements in image mapping have confirmed nanoparticles as small as 100 nm using this technique (Sobhani et al., 2020). However, strong fluorescence from additives and pigments might hinder polymer identification via Raman spectroscopy, necessitating meticulous sample purification for accurate results. Notably, Raman spectroscopy is not affected by sample thickness or moisture, distinguishing it from FTIR. While automation is feasible, handling larger sample volumes remains a challenge, similar to other spectroscopic methods discussed.

11.10 MPS IN AQUATIC ENVIRONMENTS

11.10.1 MARINE SYSTEMS

MP pollution is a significant concern in marine environments (Ma et al., 2016). These small plastic particles are widely distributed across the world's oceans and seas, from the Atlantic to the Pacific Ocean, and from the Caribbean to the Mediterranean Sea. Recently, MPs have been found in previously unexpected places like Arctic sea ice, the Antarctic, remote mountain ranges, and deep ocean trenches (Waller et al., 2017). Around 80% of marine plastic waste originates from inland sources, predominantly transported to the oceans through rivers (Mani et al., 2016). The contamination of marine environments with MPs hinges on several factors, including natural elements like wind currents and coastline geology, as well as human-induced factors such as poorly managed plastic waste disposal and uncontrolled industrial emissions (Barnes et al., 2009). Interestingly, these MPs can even evade water treatment plant processes (Fok et al., 2017). Moreover, they can travel via inland streams to estuaries and eventually reach the marine environment (Lechner et al., 2014; Rech et al., 2020). In a study by Selvam et al. (2020a), it was discovered that the Punnakayal estuary, located on the southeast coast of India, was contaminated with up to 19.9 MP particles per liter. This indicates the estuary's role in transporting MPs from inland sources to the Gulf of Mannar (Selvam et al., 2020a). Another study by Manickavasagam et al. (2020) estimated the transportation of plastic debris from densely populated areas to seas via the South Juhu creek (Manickavasagam et al., 2020). This study revealed that a substantial portion of transported plastic debris consisted of larger plastic pieces that fragment over time, ultimately converting into MPs, highlighting a critical issue that needs addressing. Due to the ongoing mismanagement of plastic waste, predictions indicate that by 2050, the amount of MPs in our oceans will surpass the total number of fish (World Economic Forum, 2016).

Ajith et al. (2020) analyzed the global contribution of various geographical areas to marine MP debris, finding that most of it originates from primary sources. Their findings showcased a 15.9% contribution from Southeast Asia, 17.2% from North America, and 8.7% from Africa and the Middle East. While researchers have conducted numerous expeditions to estimate MP abundance in regions such as the Pacific Ocean, Atlantic Ocean, Bay of Bengal, and Southern Ocean, the Indian Ocean has received relatively less attention and exploration in this regard (Ajith et al., 2020).

11.10.2 MARINE SEDIMENTS

MPs begin to accumulate in benthic layers, particularly sediments, when their density exceeds that of seawater (> 1020 kg/m^3); otherwise, they remain floating on the surface (Cauwenberghe et al., 2015). Sediments are recognized as long-term repositories for MPs, given that floating particles are easier to remove compared to those embedded in sediment layers (Lima et al., 2014). Reports extensively document the presence of MPs in marine sediments across various regions in India, including Gujarat, Tamil Nadu, Goa, Pondicherry, Maharashtra, Kerala, Karnataka, Andaman and Nicobar Islands, and Lakshadweep. The pioneering study investigating MP contamination in India's marine environments was conducted by Reddy et al. (2006), with a focus on intertidal sediments at a ship-breaking yard in Gujarat. Characterization of MPs using FTIR spectroscopy identified nylon, polyester, polyurethane, and PS as the predominant polymers present in the sediments, commonly used in ship construction (Reddy et al., 2006).

Beaches play a critical role as repositories for highly fragmented plastic debris and can act as pathways for transporting these MPs into coastal waters (Fok et al., 2017). The presence of MPs on these beaches stems from various natural and human-induced sources. Human-induced sources encompass activities such as fishing, tourism, recreation, religious practices, port operations, industrial activities, poorly managed plastic waste, and untreated wastewater discharges. On the other hand, natural sources include processes like surface and wind currents, aeolian actions, runoff, and river transportation (Jayasiri et al., 2013b, Balasubramaniam and Phillott, 2016; Tiwari et al., 2019; Maharana et al., 2020; Robin et al., 2020). Numerous studies indicate that irregularly shaped fragments comprise a significant majority of the MPs found on these beaches. Furthermore, these studies suggest that the level of natural and human activities dictates the abundance of MPs in a specific area. For example, a study across six beaches in Puducherry revealed a noteworthy correlation between MPs and the intensity of tourism activities (Dowarah et al., 2020). In a similar study, Karthik et al. (2018) examined 25 beaches along Tamil Nadu's south coast and discovered that beaches close to rivers had a higher abundance of MPs. This finding indicates that rivers are the primary transporters of these particles from land-based sources (Karthik et al., 2018).

11.10.3 MARINE ORGANISMS

India has a significant population that depends on coastal and marine ecosystems and their resources (Kumar et al., 2018). Despite this dependence, these ecosystems face continuous threats from various human activities. Among these threats, MPs are emerging as a relatively understudied yet significant concern for marine organisms. Current understanding suggests that a considerable number of organisms are susceptible to ingesting these synthetic polymers, leading to various health complications. Studies indicate that exposure to MPs can induce oxidative stress, decrease filtration capacity, cause tissue inflammation, impair the digestive tract, trigger pseudo-satiation, and reduce immunity in marine organisms (Daniel et al., 2020; Maharana et al., 2020; Dowarah et al., 2020; Sathish et al., 2020b).

In India, researchers have been particularly interested in studying the ingestion of MPs by fishes. Studies have found MPs in fish ranging from 0.05 to 10.65 particles per individual, while in marine waters, the concentration ranges from 0.93 to 126.6 x10^3 particles per cubic meter. Recent studies on the impact of MPs on a marine fish species, Carassius carassius, showed that ingesting 15–76 particles per individual over a 6-week exposure period led to weight loss, disruptions in the buccal cavity, liver inflammation, and the formation of microgranulomas (Jabeen et al., 2018; Wang et al., 2020a). Additionally, a concentration of 100 MP particles per liter over a 96-hour exposure period resulted in reduced predatory performance and overall efficiency in Pomatoschistus microps, based on habitat abundance (Wang et al., 2020a).

11.11 CONCLUSION

In conclusion, the pervasive issue of MP pollution in aquatic environments demands urgent attention and comprehensive action. The detrimental impact of MPs on marine life, ecosystems, and even human health through the consumption of contaminated seafood underscores the need for effective strategies to mitigate this environmental menace. Understanding the various sources and classifications of MPs provides a foundation for targeted interventions. Primary and secondary MPs, originating from activities such as cleaning processes, textile laundering, personal care products, paints, and packaging materials, contribute significantly to the pollution of freshwater and marine systems. Single-use plastics and domestic items, in particular, emerge as significant contributors to marine pollution, emphasizing the role of individual consumption habits in exacerbating this crisis. The factors influencing the abundance of MPs, including population density, urban proximity, and the impact of rainfall on runoff, highlight the complex interplay between human activities and environmental dynamics.

Efforts to address these factors must consider both land-based and marine-related sources, adopting a holistic approach to tackle the multifaceted nature of MP pollution. The discrepancy in findings regarding the contribution of sewage effluents to MP pollution underscores the need for standardized assessment methods and improves wastewater treatment technologies. Achieving a better understanding of the distribution, size, and polymer composition of MPs in wastewater effluents is crucial for designing effective and targeted treatment strategies. In conclusion, combating MP pollution requires a combination of regulatory measures, public awareness campaigns, and technological innovations. Sustainable practices, such as reducing single-use plastic consumption, promoting responsible waste disposal, and investing in eco-friendly alternatives, are pivotal in safeguarding aquatic ecosystems for future generations. As we navigate these challenges, collaboration among scientists, policymakers, industries, and the public is essential to enact lasting change and preserve the health and biodiversity of our oceans. In conclusion, the pervasive presence of MPs in aquatic environments poses multifaceted challenges, encompassing ecological, human health, and environmental dimensions. The choice of sampling techniques and analytical methods plays a pivotal role in accurately estimating the abundance of MPs, and it is crucial to consider potential sources of contamination during the sampling process. The distribution and movement of MPs in both freshwater and marine ecosystems are influenced by various factors such as water currents, flow rates, and environmental conditions. Lakes, rivers, and estuaries act as reservoirs and conduits for MPs, with urban areas serving as significant contributors, particularly through stormwater runoff. The risk of groundwater pollution due to MPs driven deeper into subsurface layers by freeze-thaw cycles underscores the complex interactions between environmental processes and MP fate. WWTPs play a crucial role in mitigating MP pollution, yet the efficiency of removal varies, and biosolids from treatment processes may concentrate MPs, raising concerns about their recycling and potential entry into the food chain. Bioaccumulation and biomagnification of MPs in aquatic ecosystems have far-reaching consequences for both wildlife and human populations. The potential for MPs to adsorb and transport toxic pollutants within the food chain amplifies the risks associated with their ingestion. The effects on aquatic life, especially fish, include tissue damage, oxidative stress, altered immune response, neurotoxicity, and behavioral abnormalities. The interconnectedness of environmental compartments, including freshwater, marine, terrestrial, and atmospheric environments, emphasizes the need for a holistic approach to address the widespread issue of MP pollution. Collaborative efforts among researchers, policymakers, industries, and communities are essential to develop effective mitigation strategies, enhance wastewater treatment technologies, and promote sustainable practices that reduce the production and release of MPs into the environment. As we navigate the complex web of MP interactions, a proactive and informed approach is imperative to safeguard the health of ecosystems, wildlife, and ultimately, human well-being.

KEYWORDS

- **bioaccumulation**
- **ecological impact**
- **human exposure**
- **freshwater systems**
- **microplastic pollution**

REFERENCES

Ajith, N., Arumugam, S., Parthasarathy, S., Manupoori, S., & Janakiraman, S. (2020). Global distribution of microplastics and its impact on marine environment—A review. *Environmental Science and Pollution Research International*, *27*, 25970–25986. https://doi.org/10.1007/s11356-020-09015-5

Akhbarizadeh, R., Moore, F., & Keshavarzi, B. (2019). Investigating microplastics bioaccumulation and biomagnification in seafood from the Persian Gulf: A threat to human health. *Food Additives & Contaminants: Part A*, *36*(11), 1696–1708.

Alexander, D. E. (1999). *Bioaccumulation, Bioconcentration, Biomagnification*. Springer.

Alimi, O. S., Fadare, O. O., & Okoffo, E. D. (2020). Microplastics in African ecosystems: Current knowledge, abundance, associated contaminants, techniques, and research needs. *Science of The Total Environment*, *755*, 142422.

Al-Thawadi, S. (2020). Microplastics and nanoplastics in aquatic environments: Challenges and threats to aquatic organisms. *Arabian Journal for Science and Engineering*, *45*, 4419–4440. https://doi.org/10.1007/s13369-020-04402-z

an Cauwenberghe, L., Vanreusel, A., Mees, J., & Janssen, C. R. (2013). Microplastic pollution in deep-sea sediments. *Environmental Pollution*, *182*, 495–499.

Andrady, A. L. (2011). Microplastics in the marine environment. *Marine Pollution Bulletin*, *62*(8), 1596–1605.

Arthur, C., Baker, J. E., & Bamford, H. A. (2009). *Proceedings of the International Research Workshop on the Occurrence, Effects, and Fate of Microplastic Marine Debris, September 9–11, 2008*, University of Washington Tacoma, Tacoma, WA, USA.

Auta, H. S., Emenike, C. U., & Fauziah, S. H. (2017). Distribution and importance of microplastics in the marine environment: A review of the sources, fate, effects, and potential solutions. *Environment International*, *102*, 165–176. https://doi.org/10.1016/j.envint.2017.02.013

Bai, Z., Wang, N., & Wang, M. (2021). Effects of microplastics on marine copepods. *Ecotoxicology and Environmental Safety*, 217, 112243.

Balasubramaniam, M., & Phillott, A. D. (2016). Preliminary observations of microplastics from beaches of Indian Ocean. In: *Indian Ocean Turtle Newsletter No. 23*, pp. 13–16. http://seaturtle.org/library/BalasubramaniamM_2016_IndOceanTurtleNewsletter.pdf.

Barnes, D. K. A., Galgani, F., Thompson, R. C., & Barlaz, M. (2009). Accumulation and fragmentation of plastic debris in global environments. *Philosophical Transactions of the Royal Society of London. Series B, Biological Sciences*, *364*, 1985–1998. https://doi.org/10.1098/rstb.2008.0205

Bergmann, M., Mützel, S., Primpke, S., Tekman, M. B., Trachsel, J., & Gerdts, G. (2019). White and wonderful? Microplastics prevail in snow from the Alps to the Arctic. *Science Advances*, *5*, eaax1157. https://doi.org/10.1126/sciadv.aax1157

Boerger, C. M., Lattin, G. L., Moore, S. L., & Moore, C. J. (2010). Plastic ingestion by planktivorous Fishes in the north pacific central gyre. *Marine Pollution Bulletin*, *60*(12), 2275–2278. https://doi.org/10.1016/j.marpolbul.2010.08.007.

Borrelle, S. B., Rochman, C. M., Liboiron, M., Bond, A. L., Lusher, A., Bradshaw, H., et al. (2017). Opinion: Why we need an international agreement on marine plastic pollution. *Proceedings National Academy of Science USA*, *114*(38), 9994–9997. https://doi.org/10.1073/pnas.1714450114

Browne, M. A., Crump, P., Niven, S. J., Teuten, E., Tonkin, A., Galloway, T., & Thompson, R. (2011). Accumulation of microplastic on shorelines worldwide: Sources and sinks. *Environmental Science & Technology*, *45*(21), 9175–9179.

Browne, M. A., Galloway, T. S., & Thompson, R. C. (2010). Spatial patterns of plastic debris along estuarine shorelines. *Environmental Science & Technology*, *44*(9), 3404–3409.

Cai, L., Wang, J., Peng, J., Tan, Z., Zhan, Z., Tan, X., & Chen, Q. (2017). Characteristic of microplastics in the atmospheric fallout from Dongguan city, China: Preliminary research and first evidence. *Environmental Science and Pollution Research*, *24*, 24928–24935.

Carr, S. A., Liu, J., & Tesoro, A. G. (2016). Transport and fate of microplastic particles in wastewater treatment plants. *Water Research*, *91*, 174–182.

Cauwenberghe, L. V., Devriese, L., Galgani, F., Robbens, J., & Janssen, C. R. (2015). Microplastics in sediments: A review of techniques, occurrence and effects. *Marine Environment Research*, *111*, 5–17. https://doi.org/10.1016/j.marenvres.2015.06.007

Claessens, M., De Meester, S., Van Landuyt, L., De Clerck, K., & Janssen, C. R. (2011). Occurrence and distribution of microplastics in marine sediments along the Belgian coast. *Marine Pollution Bulletin*, *62*(10), 2199–2204.

Corcoran, P. L., Norris, T., Ceccanese, T., Walzak, M. J., Helm, P. A., & Marvin, C. H. (2015). Hidden plastics of Lake Ontario, Canada and their potential preservation in the sediment record. *Environmental Pollution, 204*, 17–25.

Crawford, C. B., & Quinn, B. (2016). *Microplastic Pollutants*. Elsevier.

Daniel, D. B., Ashraf, P. M., & Thomas, S. N. (2020). Abundance, characteristics and seasonal variation of microplastics in Indian white shrimps (*Fenneropenaeus indicus*) from coastal waters off Cochin, Kerala, India. *Science of the Total Environment, 737*, 139839.

Dawson, A., Huston, W., Kawaguchi, S., King, C., Cropp, R., Wild, S., Eisenmann, P., Townsend, K., & Bengtson Nash, S. (2018). Uptake and depuration kinetics influence microplastic bioaccumulation and toxicity in Antarctic krill (*Euphausia superba*). *Environmental Science & Technology, 52*(5), 3195–3201.

de Sá, L. C., Luís, L. G., & Guilhermino, L. (2015). Effects of MPs on juveniles of the common goby (*Pomatoschistus microps*): Confusion with prey, reduction of the predatory performance and efficiency, and possible influence of developmental conditions. *Environmental Pollution, 196*, 359–362.

Desforges, J. P. W., Galbraith, M., Dangerfield, N., & Ross, P. S. (2014). Widespread distribution of microplastics in subsurface seawater in the NE Pacific Ocean. *Marine Pollution Bulletin, 79*(1–2), 94–99.

Dowarah, K., Patchaiyappan, A., Thirunavukkarasu, C., Jayakumar, S., & Devipriya, S. P. (2020). Quantification of microplastics using Nile red in two bivalve species *Pernaviridis* and *Meretrixmeretrix* from three estuaries in Pondicherry, India and microplastic uptake by local communities through bivalve diet. *Marine Pollution Bulletin, 153*, 110982. https://doi.org/10.1016/j.marpolbul.2020.110982

Dris, R., Gasperi, J., Mirande, C., Mandin, C., Guerrouache, M., Langlois, V., & Tassin, B. (2017). A first overview of textile fibers, including microplastics, in indoor and outdoor environments. *Environmental Pollution, 221*, 453–458.

Dris, R., Gasperi, J., Saad, M., Mirande, C., & Tassin, B. (2016). Synthetic fibers in atmospheric fallout: A source of microplastics in the environment. *Marine Pollution Bulletin, 104*(1–2), 290–293.

Duis, K., & Coors, A. (2016). Microplastics in the aquatic and terrestrial environment: Sources (with a specific focus on personal care products), fate and effects. *Environmental Sciences Europe, 28*(1), 1–25.

FAO. (2016). The State of World Fisheries and Aquaculture 2016. *Contributing to Food Security and Nutrition for All*. Rome, p. 200.

Farrell, P., & Nelson, K. (2013). Trophic level transfer of microplastic: Mytilusedulis (L.) to Carcinusmaenas (L.). *Environmental Pollution, 177*, 1–3.

Fisheries of the United States. (2015). Fisheries of the United States.

Fok, L., Cheung, P. K., Tang, G., & Li, W. C. (2017). Size distribution of stranded small plastic debris on the coast of Guangdong, South China. *Environmental Pollution, 220*, 407–412. https://doi.org/10.1016/j.envpol.2016.09.079

Free, C. M., Jensen, O. P., Mason, S. A., Eriksen, M., Williamson, N. J., & Boldgiv, B. (2014). High-levels of microplastic pollution in a large, remote, mountain lake. *Marine Pollution Bulletin*, 85(1), 156–163.

Fries, E., Dekiff, J. H., Willmeyer, J., Nuelle, M. T., Ebert, M., & Remy, D. (2013). Identification of polymer types and additives in marine microplastic particles using pyrolysis-GC/MS and scanning electron microscopy. *Environmental Science: Processes & Impacts, 15*(10), 1949–1956.

Fu, W., Min, J., Jiang, W., Li, Y., & Zhang, W. (2020). Separation, characterization and identification of microplastics and nanoplastics in the environment. *Science of the Total Environment*, *721*, 137561.

Gerber, K. (2009). User's guide and technical documentation KABAM version 1.0 (Kow (based) Aquatic Bioaccumulation Model). *Report of Environmental Fate and Effects Division Office of Pesticide Programs, US Environmental Protection Agency*, Washington, DC.

Guo, J. J., Huang, X. P., Xiang, L., Wang, Y. Z., Li, Y. W., Li, H., Cai, Q. Y., Mo, C. H., & Wong, M. H. (2020). Source, migration and toxicology of microplastics in soil. *Environment International*, *137*, 105263.

Gurjar, U. R., Xavier, M., Nayak, B. B., Ramteke, K., Deshmukhe, G., Jaiswar, A. K., & Shukla, S. P. (2021). Microplastics in shrimps: A study from the trawling grounds of north eastern part of Arabian Sea. *Environmental Science and Pollution Research,* 28(35), 48494–48504.

Harrison, J. P., Ojeda, J. J., & Romero-González, M. E. (2012). The applicability of reflectance micro-Fourier-transform infrared spectroscopy for the detection of synthetic microplastics in marine sediments. *Science of the Total Environment*, *416*, 455–463.

Hidalgo-Ruz, V., Gutow, L., Thompson, R. C., & Thiel, M. (2012). Microplastics in the marine environment: A review of the methods used for identification and quantification. *Environmental Science & Technology, 46*(6), 3060–3075.

Huang, J.-N., Wen, B., Xu, L., Ma, H.-C., Li, X.-X., Gao, J.-Z., et al. (2022). Micro/nano-plastics cause neurobehavioral toxicity in discus fish (*Symphysodon aequifasciatus*): Insight from brain-gut-microbiota axis. *Journal of Hazardous Materials*, *421*, 126830. https://doi.org/10.1016/j.jhazmat.2021.126830

Jabeen, K., Li, B., Chen, Q., Su, L., Wu, C., Hollert, H., et al. (2018). Effects of virgin MPs on Goldfish (Carassiusauratus). *Chemosphere*, *213*, 323–332. https://doi.org/10.1016/j.chemosphere.2018.09.031

Jang, M., Shim, W. J., Cho, Y., Han, G. M., Song, Y. K., & Hong, S. H. (2020). A close relationship between microplastic contamination and coastal area use pattern. *Water Research*, *171*, 115400.

Jayasiri, H. B., Purushothaman, C. S., & Vennila, A. (2013b) Quantitative analysis of plastic debris on recreational beaches in Mumbai, India. *Marine Pollution Bulletin*, *77*, 107–112. https://doi.org/10.1016/j.marpolbul.2013.10.024

Karami, A., Golieskardi, A., KeongChoo, C., Larat, V., Galloway, T. S., & Salamatinia, B. (2017). The presence of microplastics in commercial salts from different countries. *Scientific Reports*, *7*(1), 46173.

Karthik, R., Robin, R. S., Purvaja, R., Ganguly, D., Anandavelu, I., Raghuraman, R., Hariharan, G., Ramakrishna, A., & Ramesh, R. (2018). Microplastics along the beaches of southeast coast of India. *Science of the Total Environment*, *645*, 1388–1399.

Kelly, B. C., Ikonomou, M. G., Blair, J. D., Morin, A. E., & Gobas, F. A. (2007). Food web specific biomagnification of persistent organic pollutants. *Science*, *317*(5835), 236–239.

Klein, M., & Fischer, E. K. (2019). Microplastic abundance in atmospheric deposition within the Metropolitan area of Hamburg, Germany. *Science of the Total Environment*, *685*, 96–103.

Koutnik, V. S., Leonard, J., Brar, J., Cao, S., Glasman, J. B., Cowger, W., Ravi, S., & Mohanty, S. K. (2022). Transport of microplastics in stormwater treatment systems under freeze-thaw cycles: Critical role of plastic density. *Water Research, 222*, 118950. https://doi.org/10.1016/j.watres.2022.118950

Kumar, V. E., Ravikumar, G., & Jeyasanta, K. I. (2018). Occurrence of microplastics in fishes from two landing sites in Tuticorin, south east coast of India. *Marine Pollution Bulletin*, *135*, 889–894. https://doi.org/10.1016/j.marpolbul.2018.08.023

Kumkar, P., Gosavi, S. M., Verma, C. R., Pise, M., & Kalous, L. (2021). Big eyes can't see MPs: Feeding selectivity and eco-morphological adaptations in oral cavity affect microplastic uptake in mud-dwelling amphibious mudskipper fish. *Science of the Total Environment*, 786, 147445. https://doi.org/10.1016/j.scitotenv.2021.147445

Lechner, A., Keckeis, H., Lumesberger-Loisl, F., Zens, B., Krusch, R., Tritthart, M., Glas, M., & Schludermann, E. (2014). The Danube so colourful: A potpourri of plastic litter outnumbers fish larvae in Europe's second largest river. *Environmental Pollution*, *188*, 177–181. https://doi.org/10.1016/j.envpol.2014.02.006

Lee, E., & O'Donnell, B. (2019). Characterizing microplastic fibers using Raman spectroscopy.

Lee, K. W., Shim, W. J., Kwon, O. Y., & Kang, J. H. (2013). Size-dependent effects of micro polystyrene particles in the marine copepod *Tigriopus japonicus*. *Environmental Science & Technology*, *47*(19), 11278–11283.

Li, J., Yang, D., Li, L., Jabeen, K., & Shi, H. (2015). Microplastics in commercial bivalves from China. *Environmental Pollution, 207*, 190–195.

Lima, A. R. A., Costa, M. F., & Barletta, M. (2014). Distribution patterns of microplastics within the plankton of a tropical estuary. *Environmental Research*, *132*, 146–155. https://doi.org/10.1016/j.envres.2014.03.031

Lindeque, P. K., Cole, M., Coppock, R. L., Lewis, C. N., Miller, R. Z., Watts, A. J., Wilson-McNeal, A., Wright, S. L., & Galloway, T. S. (2020). Are we underestimating microplastic abundance in the marine environment? A comparison of microplastic capture with nets of different mesh-size. *Environmental Pollution*, *265*, 114721.

Lu, Y., Zhang, Y., Deng, Y., Jiang, W., Zhao, Y., Geng, J., Ding, L., & Ren, H. (2016). Uptake and accumulation of polystyrene microplastics in zebrafish (Daniorerio) and toxic effects in liver. *Environmental Science & Technology*, *50*(7), 4054–4060.

Lusher, A. L., Mchugh, M., & Thompson, R. C. (2013). Occurrence of MPs in the gastrointestinal tract of pelagic and demersal fish from the English channel. *Marine Pollution Bulletin*, *67*(1–2), 94–99. https://doi.org/10.1016/j.marpolbul.2012.11.028

Lusher, A. L., Welden, N. A., Sobral, P., & Cole, M. (2020). Sampling, isolating and identifying microplastics ingested by fish and invertebrates. *Analysis of Nanoplastics and Microplastics in Food*, 119–148. https://doi.org/10.1201/9780429469596-8

Lusher, A., Hollman, P., & Mendoza-Hill, J. (2017). *Microplastics in Fisheries and Aquaculture: Status of Knowledge on Their Occurrence and Implications for Aquatic Organisms and Food Safety*. FAO.

Ma, Y., Huang, A., Cao, S., Sun, F., Wang, L., Guo, H., & Ji, R. (2016). Effects of nanoplastics and microplastics on toxicity, bioaccumulation, and environmental fate of phenanthrene in fresh water. *Environmental Pollution*, *219*, 166–173. https://doi.org/10.1016/j.envpol.2016.10.061

Mackay, D., & Boethling, R. S. (Eds.). (2000). *Handbook of Property Estimation Methods for Chemicals: Environmental Health Sciences*. CRC Press.

Maharana, D., Saha, M., Dar, J. Y., Rathore, C., Sreepada, R. A., Xu, X.-R., Koongolla, J. B., & Li, H.-X. (2020). Assessment of plastics along the west coast of India: Abundance, distribution, polymer type and toxicity. *Chemosphere*, *246*, 125708. https://doi.org/10.1016/j.chemosphere.2019.125708

Mammo, F., Amoah, I., Gani, K., Pillay, L., Ratha, S., Bux, F., et al. (2020). Microplastics in the environment: Interactions with microbes and chemical contaminants. *Science of the Total Environment*, *743*, 140518. https://doi.org/10.1016/j.scitotenv.2020.140518

Mani, T., Hauk, A., Walter, U., & Burkhardt-Holm, P. (2016). Microplastics profile along the Rhine River. *Scientific Reports*, *5*, 17988. https://doi.org/10.1038/srep17988

Manickavasagam, S., Kumar, S., Kumar, K., Bhuvaneswari, G. R., Paul, T., & Shukla, S. P. (2020). Quantitative assessment of influx and efflux of marine debris in a water channel of South Juhu creek, Mumbai. *India Regional Studies in Marine Science*, *34*, 101095. https://doi.org/10.1016/j.rsma.2020.101095

Mattsson, K., Hansson, L., & Cedervall, T. (2015). Nano-plastics in the aquatic environment. *Environmental Science: Processes & Impacts, 17*(1), 1712–1721.

MCole, M., Lindeque, P., Fileman, E., Halsband, C., Goodhead, R., Moger, J., & Galloway, T. S. (2013). Microplastic ingestion by zooplankton. *Environmental Science & Technology*, *47*(12), 6646–6655.

Mintenig, S. M., Löder, M. G., Primpke, S., & Gerdts, G. (2019). Low numbers of microplastics detected in drinking water from ground water sources. *Science of the Total Environment*, *648*, 631–635.

Mizraji, R., Ahrendt, C., Perez-Venegas, D., Vargas, J., Pulgar, J., Aldana, M., et al. (2017). Is the feeding type related with the content of MPs in intertidal fish gut? *Marine Pollution Bulletin*, *116*(1–2), 498–500. https://doi.org/10.1016/j.marpolbul.2017.01.008

Mu, X., Qi, S., Liu, J., Yuan, L., Huang, Y., Xue, J., et al. (2021). Toxicity and behavioral response of Zebrafish exposed to combined microplastic and bisphenol analogues. *Environmental Chemistry Letters*, *20*, 41–48. https://doi.org/10.1007/s10311-021-01320-w

Murphy, F., Russell, M., Ewins, C., & Quinn, B. (2017). The uptake of microplastic & microplastic by demersal & pelagic fish in the Northeast Atlantic around Scotland. *Marine Pollution Bulletin*, *122*(1–2), 353–359.

Murray, F., & Cowie, P. R. (2011). Plastic contamination in the decapod crustacean *Nephrops norvegicus* (Linnaeus, 1758). *Marine Pollution Bulletin*, *62*(6), 1207–1217.

Nor, N. H. M., & Obbard, J. P. (2014). Microplastics in Singapore's coastal mangrove ecosystems. *Marine Pollution Bulletin*, *79*(1–2), 278–283.

Pazos, R. S., Maiztegui, T., Colautti, D. C., Paracampo, A. H., & Gómez, N. (2017). MPs in gut contents of coastal freshwater fish from Río de la Plata estuary. *Marine Pollution Bulletin*, *122*(1–2), 85–90. https://doi.org/10.1016/j.marpolbul.2017.06.007

Peda, C., Caccamo, L., Fossi, M. C., Gai, F., Andaloro, F., Genovese, L., et al. (2016). Intestinal alterations in European sea bass *Dicentrarchus labrax* (Linnaeus, 1758) exposed to MPs: Preliminary results. *Environmental Pollution*, *212*, 251–256. https://doi.org/10.1016/j.envpol.2016.01.083

Petersen, F., & Hubbart, J. A. (2021). The occurrence and transport of microplastics: The state of the science. *Science of the Total Environment*, *758*, 143936.

Prajapati, S., Beal, M., Maley, J., & Brinkmann, M. (2021). Qualitative and quantitative analysis of microplastics and microfiber contamination in effluents of the City of Saskatoon wastewater treatment plant. *Environmental Science and Pollution Research*, *28*, 32545–32553.

Ragoobur, D., Huerta-Lwanga, E., & Somaroo, G. D. (2021). Microplastics in agricultural soils, wastewater effluents and sewage sludge in Mauritius. *Science of the Total Environment*, *798*, 149326.

Rech, S., Macaya-Caquilpán, V., Pantoja, J. F., Rivadeneira, M. M., Madariaga, D. J., & Thiel, M. (2020). Rivers as a source of marine litter-a study from the SE Pacific. *Marine Pollution Bulletin*, *82*, 66–75. https://doi.org/10.1016/j.marpolbul.2014.03.019

Reddy, M. S., Basha, S., Adimurthy, S., & Ramachandraiah, G. (2006). Description of the small plastics fragments in marine sediments along the Alang-Sosiya ship-breaking yard, India. *Estuarine, Coastal and Shelf Science*, *68*, 656–660. https://doi.org/10.1016/j.ecss.2006.03.018

Robin, R. S., Karthik, R., Purvaja, R., Ganguly, D., Anandavelu, I., Mugilarasan, M., & Ramesh, R. (2020). Holistic assessment of microplastics in various coastal environmental matrices, southwest coast of India. *Science of the Total Environment*, *703*, 134947. https://doi.org/10.1016/j.scitotenv.2019.134947

Rocha-Santos, T., & Duarte, A. C. (2015). A critical overview of the analytical approaches to the occurrence, the fate and the behavior of microplastics in the environment. *TrAC Trends in Analytical Chemistry*, *65*, 47–53.

Rochman, C. M., Brookson, C., Bikker, J., Djuric, N., Earn, A., Bucci, K., Athey, S., Huntington, A., McIlwraith, H., Munno, K., De Frond, H., Kolomijeca, A., Erdle, L., Grbic, J., Bayoumi, M., Borrelle, S. B., Wu, T., Santoro, S., Werbowski, L. M., Zhu, X., & Hung, C. (2019). Rethinking microplastics as a diverse contaminant suite. *Environmental Toxicology and Chemistry*, *38*(4), 703–711.

Rochman, C. M., Tahir, A., Williams, S. L., Baxa, D. V., Lam, R., Miller, J. T., Teh, F. C., Werorilangi, S., & Teh, S. J. (2015). Anthropogenic debris in seafood: Plastic debris and fibers from textiles in fish and bivalves sold for human consumption. *Scientific Reports*, *5*(1), 14340.

Sathish, M. N., Jeyasanta, I., & Patterson, J. (2020b) Occurrence of microplastics in epipelagic and mesopelagic fishes from Tuticorin, southeast coast of India. *Science of the Total Environment*, *720*, 137614. https://doi.org/10.1016/j.scitotenv.2020.137614

Schmaltz, E., Melvin, E. C., Diana, Z., Gunady, E. F., Rittschof, D., Somarelli, J. A., Virdin, J., & Dunphy-Daly, M. M. (2020). Plastic pollution solutions: Emerging technologies to prevent and collect marine plastic pollution. *Environment International*, *144*, 106067.

Selvam, S., Jesuraja, K., Venkatramamam, S., Roy, P. D., & Kumari, V. J. (2020a) Hazardous microplastic characteristics and its role as heavy metal in groundwater and surface water of coastal south India. *Journal of Hazardous Materials*, *402*, 123786. https://doi.org/10.1016/j.jhazmat.2020.123786

Sequeira, I. F., Prata, J. C., da Costa, J. P., Duarte, A. C., & Rocha-Santos, T. (2020). Worldwide contamination of fish with microplastics: A brief global overview. *Marine Pollution Bulletin*, *160*, 111681. https://doi.org/10.1016/j.marpolbul.2020.111681

Simpson, S., & Batley, G. (Eds.). (2016). *Sediment Quality Assessment: A Practical Guide*. CSIRO Publishing.

Smith, E., & Dent, G. (2019). *Modern Raman Spectroscopy: A Practical Approach*. John Wiley & Sons.

Sobhani, Z., Zhang, X., Gibson, C., Naidu, R., Megharaj, M., & Fang, C. (2020). Identification and visualisation of microplastics/nanoplastics by Raman imaging: Down to 100 nm. *Water Research*, *174*, 115658.

Steer, M., Cole, M., Thompson, R. C., & Lindeque, P. K. (2017). Microplastic ingestion in fish larvae in the western English Channel. *Environmental Pollution*, *226*, 250–259.

Thompson, R. C., Olsen, Y., Mitchell, R. P., Davis, A., Rowland, S. J., John, A. W., McGonigle, D., & Russell, A. E. (2004). Lost at sea: Where is all the plastic. *Science*, *304*(5672), 838–838.

Thushari, G. G. N., & Senevirathna, J. D. M. (2020). Plastic pollution in the marine environment. *Heliyon*, *6*(8), e04709.

Tiwari, M., Rathod, T. D., Ajmal, P. Y., Bhangare, R. C., & Sahu, S. K. (2019). Distribution and characterization of microplastics in beach sand from three different Indian coastal environments. *Marine Pollution Bulletin*, *140*, 262–273. https://doi.org/10.1016/j.marpolbul.2019.01.055

Van Cauwenberghe, L., & Janssen, C. R. (2014). Microplastics in bivalves cultured for human consumption. *Environmental Pollution*, *193*, 65–70.

Vianello, A., Boldrin, A., Guerriero, P., Moschino, V., Rella, R., Sturaro, A., & Da Ros, L. (2013). Microplastic particles in sediments of Lagoon of Venice, Italy: First observations on occurrence, spatial patterns and identification. *Estuarine, Coastal and Shelf Science*, *130*, 54–61.

Waller, C. L., Griffiths, H. J., Waluda, C. M., Thorpe, S. E., Loaiza, I., Moreno, B., Pacherres, C. O., & Hughes, K. H. (2017). Microplastics in the Antarctic marine system: An emerging area of research. *Science of the Total Environment*, *598*, 220–227. https://doi.org/10.1016/j.scitotenv.2017.03.283

Wang, J., Tan, Z., Peng, J., Qiu, Q., & Li, M. (2016). The behaviors of microplastics in the marine environment. *Marine Environmental Research*, *113*, 7–17.

Wang, W., & Wang, J. (2018). Investigation of microplastics in aquatic environments: An overview of the methods used, from field sampling to laboratory analysis. *TrAC Trends in Analytical Chemistry*, *108*, 195–202.

Wang, W., Ge, J., & Yu, X. (2020a) Bioavailability and toxicity of microplastics to fish species: A review. *Ecotoxicology and Environmental Safety*, *189*, 109913. https://doi.org/10.1016/j.ecoenv.2019.109913

Weber, W. H., & Merlin, R. (Eds.). (2000). *Raman Scattering in Materials Science*. Springer Science & Business Media, Vol. 42.

Wesch, C., Bredimus, K., Paulus, M., & Klein, R. (2016). Towards the suitable monitoring of ingestion of MPs by marine Biota: *A Review. Environmental Pollution*, *218*, 1200–1208. https://doi.org/10.1016/j.envpol.2016.08.076.

World Economic Forum. (2016). The new plastics economy: Rethinking the future of plastics. http://www3.weforum.org/docs/WEF_The_New_Plastics_Economy.pdf. Accessed 30 November 2020.

Wright, S. L., Thompson, R. C., & Galloway, T. S. (2013). The physical impacts of MPs on marine organisms: A review. *Environmental Pollution*, *178*, 483–492. https://doi.org/10.1016/j.envpol.2013.02.031

Young, A. M., & Elliott, J. A. (2016). Characterization of microplastic and mesoplastic debris in sediments from Kamilo Beach and Kahuku Beach, Hawai'i. *Marine Pollution Bulletin*, *113*(1–2), 477–482.

Zhang, Y., Kang, S., Allen, S., Allen, D., Gao, T., & Sillanpää, M. (2020). Atmospheric microplastics: A review on current status and perspectives. *Earth-Science Reviews, 203*, 103118.

Zhang, K., Shi, H., Peng, J., Wang, Y., Xiong, X., Wu, C., & Lam, P. K. (2018). Microplastic pollution in China's inland water systems: A review of findings, methods, characteristics, effects, and management. *Science of the Total Environment*, *630*, 1641–1653.

Zhang, X., Xia, M., Zhao, J., Cao, Z., Zou, W., & Zhou, Q. (2022). Photoaging enhanced the adverse effects of polyamide MPs on the growth, intestinal health, and lipid absorption in developing Zebrafish. *Environment International*, *158*, 106922. https://doi.org/10.1016/j.envint.2021.106922

Zhao, Y., Qin, Z., Huang, Z., Bao, Z., Luo, T., & Jin, Y. (2021). Effects of polyethylene MPs on the microbiome and metabolism in larval zebrafish. *Environmental Pollution*, *282*, 117039. https://doi.org/10.1016/j.envpol.2021.117039

CHAPTER 12

The Nexus of Nanotechnology and Freshwater Aquaculture: Driving Forward with Innovation

GUNTAPALLI SRAVANI[1*] and SURYA CHAITANYA BANURU[2]

[1]*Division of Aquaculture, ICAR-CIFE, Mumbai, Maharashtra, India*

[2]*AAHM Division, ICAR-CIFE, Mumbai, Maharashtra, India*

*Corresponding author

ABSTRACT

Nanotechnology is increasingly transforming aquaculture by offering innovative solutions for disease control, water purification, and feed efficiency. Emerging in the early 21st century, it involves manipulating materials at the nanoscale (1–100 nm) to harness their unique properties. In aquaculture, nanoparticles are utilized to enhance water quality through advanced filtration systems and real-time monitoring sensors, enabling the removal of microscopic contaminants. For disease management, nanotechnology supports the development of targeted drug delivery systems and nanovaccines, offering greater treatment efficacy with fewer side effects. In nutrition, nanoparticles improve feed performance by enhancing nutrient absorption and digestion. Nanoencapsulation techniques protect essential nutrients, increasing their bioavailability and stability, ultimately improving feed conversion ratios and promoting better growth in aquatic species. While these innovations support sustainable aquaculture practices and improve fish health and productivity, concerns remain regarding the environmental safety and potential

Advances in Freshwater Aquaculture: Technology for Sustainable Development. Tarang Kumar Shah, Madhu Sharma, Mohd. Ashraf Rather, & Avdhesh Kumar (Eds.)

DOI: 10.1201/9781779640857-12

toxicity of nanoparticles. Therefore, the adoption of nanotechnology must be accompanied by robust regulatory frameworks and comprehensive risk assessments to mitigate ecological impacts and ensure responsible implementation.

12.1 INTRODUCTION

Nanoparticles are minuscule particles that reside within the nanoscale, typically measuring between 1 and 100 nm across at least one dimension (Selvaraj et al., 2014). Nanoparticles exhibit a set of unique characteristics that distinguish them from their bulk material counterparts. Due to their diminutive size, these particles exhibit distinct physical and chemical characteristics that render them valuable across various sectors, including electronics, healthcare, and materials engineering (Sravani et al., 2024). Nanoparticles are pivotal in driving advances across various fields by leveraging their unique size-dependent properties. As the field of nanotechnology expands, the diversity and applications of these nanoparticles are expected to increase significantly (Samanta et al., 2022). These special features are predominantly due to their nanoscale dimensions, which confer properties not observed in larger forms of the same materials.

12.2 KEY CHARACTERISTICS OF NANOPARTICLES

12.2.1 INCREASED SURFACE AREA TO VOLUME RATIO

One of the most defining features of nanoparticles is their significantly increased surface area relative to their volume. This enhanced surface area allows for more atoms to be exposed at the surface, which can interact with the surrounding environment (Sarkar et al., 2022). This property is particularly beneficial in catalysis, sensor technology, and drug delivery, where increased surface contact is crucial for improved reactivity and binding efficiency.

12.2.2 QUANTUM EFFECTS

Nanoparticles exhibit quantum mechanical effects that become significant at these small scales. Quantum effects can alter the optical, electrical, and magnetic behaviors of materials. For example, quantum dots have unique electronic properties such as size-tunable light emission, making them valuable for applications in light-emitting diodes, quantum computing, and medical imaging (Ogunfowora et al., 2021).

12.2.3 ENHANCED REACTIVITY

Due to their increased surface area and quantum effects, nanoparticles often show greater chemical reactivity compared with bulk materials. This is beneficial in various applications, including the faster catalytic conversion of chemicals, improved efficacy in photoreactive sunscreen ingredients like titanium dioxide, and enhanced efficiency in pollution control technologies (Noor et al., 2023).

12.2.4 UNIQUE OPTICAL PROPERTIES

Nanoparticles can absorb and scatter light differently compared with bulk materials. The color of gold nanoparticles, for instance, can vary from red to purple depending on their size, shape, and the degree of aggregation due to their localized surface plasmon resonance properties. This makes them useful in applications ranging from therapeutic treatments to biological imaging (Nasr-Eldahan et al., 2021).

12.2.5 MAGNETIC PROPERTIES

Certain nanoparticles exhibit unique magnetic properties that are not found in their bulk counterparts (Tayel et al., 2019). Superparamagnetism, observed in some metal-oxide nanoparticles like iron oxide, is one such property where nanoparticles become magnetized in the presence of an external magnetic field but revert to a nonmagnetic state once the field is removed. This property is exploited in magnetic resonance imaging (MRI) for enhanced imaging capabilities (Márquez et al., 2018).

12.2.6 MECHANICAL PROPERTIES

Nanoparticles can also exhibit unusual mechanical properties such as increased strength and resilience compared with bulk materials. This makes them particularly useful in the creation of composites for structural applications where lightweight yet strong materials are desirable (Khosravi-Katuli et al., 2017).

12.2.7 THERMAL PROPERTIES

The thermal conductivity of nanoparticles can also be significantly different from that of bulk materials. For instance, nanostructured materials like aerogels have extremely low thermal conductivity and are used as insulators in both industrial and domestic applications (Gunalan et al., 2012).

These unique properties enable nanoparticles to perform effectively in a wide range of applications, from industrial manufacturing processes to next-generation electronic devices and innovative medical therapies. Understanding and harnessing these characteristics are at the forefront of research in nanotechnology.

12.3 CATEGORIZATION OF NANOPARTICLES BASED ON THEIR MATERIAL COMPOSITION, MORPHOLOGY, AND STRUCTURE

12.3.1 METAL NANOPARTICLES

These particles are composed of metallic elements such as gold, silver, and platinum, renowned for their exceptional optical, electrical, and thermal attributes. Gold and silver nanoparticles, for instance, are utilized extensively in biomedicine for targeted drug delivery, diagnostic imaging, and biosensing, owing to their peculiar optical properties (Elechiguerra et al., 2005).

12.3.2 METAL-OXIDE NANOPARTICLES

Comprising metal cations bonded with oxygen, these nanoparticles include compounds like titanium dioxide, zinc oxide, and iron oxide (Deng and Cheng, 2003). They find applications in a variety of products ranging from cosmetics and sunscreens to paints and industrial coatings due to properties such as UV absorption (titanium dioxide and zinc oxide) and magnetic characteristics (iron oxide), which are useful in MRI and targeted drug delivery systems.

12.3.3 CERAMIC NANOPARTICLES

Constructed from inorganic nonmetallic materials, these nanoparticles are noted for their thermal resistance, chemical inertness, and mechanical strength. Silicon carbide and alumina are examples of ceramic nanoparticles employed in areas such as biomedical implants, electronics, and as catalyst supports (Shah and Mraz, 2020).

12.3.4 SEMICONDUCTOR NANOPARTICLES

Also recognized as quantum dots, these nanoparticles exhibit electronic characteristics that vary with their size. Materials like cadmium selenide and indium arsenide are popular choices (Alexander and Mewhinney, 2008). Quantum dots are instrumental in applications involving light-emitting diodes, solar cells, and biomedical imaging due to their tunable optical properties.

12.3.5 POLYMERIC NANOPARTICLES

Consisting of natural or synthetic polymers, these nanoparticles are pivotal in controlled drug release technologies. They enable precise delivery of therapeutics and are exemplified by compounds such as poly(lactic-*co*-glycolic acid) (PLGA) for pharmaceutical delivery and dendrimers used in gene therapy.

12.3.6 LIPID-BASED NANOPARTICLES

This category includes solid lipid nanoparticles and nanostructured lipid carriers, predominantly used in the pharmaceutical field for drug delivery and gene therapy. These nanoparticles encapsulate active agents, offering protection against degradation and facilitating controlled release.

12.3.7 CARBON-BASED NANOPARTICLES

Carbon-based nanoparticles have become essential tools in various industries due to their unique properties, and their application in aquaculture is particularly promising. These nanoparticles offer innovative solutions to enhance water quality, disease management, feed efficiency, and monitoring systems. This group encompasses various allotropes of carbon, such as fullerenes, carbon nanotubes (CNTs), graphene, and carbon quantum dots, each distinguished by unique properties.

12.3.7.1 FULLERENES

Fullerenes, with their hollow, spherical structure, have notable applications in drug delivery systems within aquaculture. Their ability to encapsulate various molecules makes them ideal for targeted therapy, delivering medications or nutrients directly to diseased or nutritionally deficient fish. This targeted delivery system helps reduce the dosage and side effects of medications, potentially lowering the environmental impact of pharmaceuticals in aquaculture.

12.3.7.2 CARBON NANOTUBES

CNTs are used for their robust mechanical properties and electrical conductivity, but their role in aquaculture extends to critical antibacterial applications.

12.3.7.2.1 Disease Prevention

Leveraging their antibacterial capabilities, CNTs are applied as coatings on aquaculture equipment like tanks and feeding instruments to inhibit the growth of harmful bacteria and biofilm formations.

12.3.7.2.2 Water Filtration

Thanks to their nanoscale size and extensive surface area, CNTs are effective in filtering systems where they capture and eliminate microscopic contaminants and pathogens, significantly improving water purity in aquaculture settings.

12.3.7.3 GRAPHENE

Graphene's superior electrical conductivity and mechanical strength are employed in various aquaculture applications.

12.3.7.3.1 Sensing Technology

Graphene sensors are exceptionally responsive, capable of detecting subtle changes in environmental parameters such as pH, temperature, and the presence of toxic substances, facilitating real-time environmental control.

12.3.7.3.2 Structural Reinforcement

Integrating graphene into materials used for aquaculture infrastructures, like nets and cages, enhances their durability and reduces the need for frequent maintenance.

12.3.7.4 CARBON QUANTUM DOTS

These fluorescent nanoparticles are used in imaging applications, water quality monitoring, and light-emitting devices.

12.3.8 COMPOSITE NANOPARTICLES

Constructed from two or more disparate materials, these nanoparticles combine properties to suit specific applications such as a magnetically responsive core coated with a drug-loaded polymer for magnetically guided drug delivery (Figure 12.1).

Owing to their diminutive scale, nanoparticles possess distinct physical and chemical attributes markedly different from their bulk material counterparts. These attributes include a high ratio of surface area to volume, pronounced quantum effects, and elevated reactivity, making them invaluable across diverse sectors such as medicine, electronics, environmental science, and aquaculture.

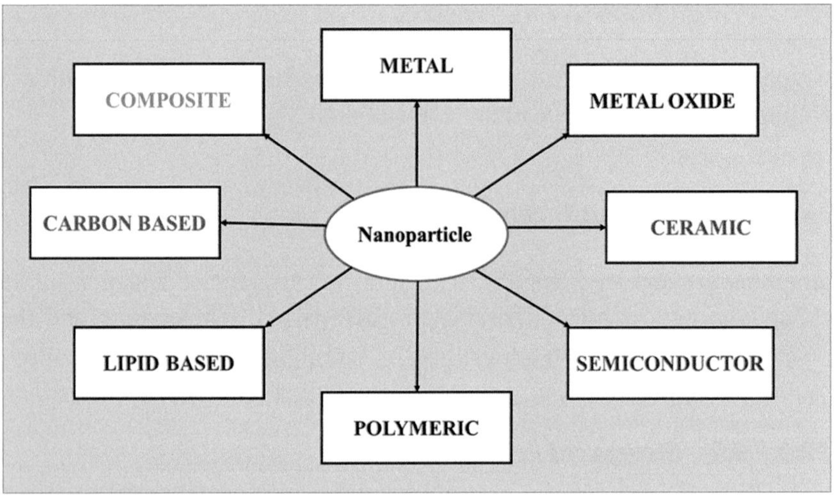

FIGURE 12.1 Different types of nanoparticles.

12.4 PRODUCTION METHODS OF NANOPARTICLES IN AQUACULTURE

The synthesis of nanoparticles for aquaculture purposes is generally divided into two principal strategies: top-down and bottom-up approaches, each with unique benefits and specific applications.

12.4.1 TOP-DOWN METHODS

These methods involve deconstructing bulk materials into finer nanoparticles using physical processes. Prominent top-down techniques are mentioned below.

12.4.1.1 BALL MILLING

This mechanical technique involves grinding a substance in a ball mill to form a fine powder (Sarkar et al., 2022). The particle size can be finely tuned by adjusting various parameters.

12.4.1.2 LASER ABLATION

In this process, a high-energy laser is used to vaporize a target material, resulting in the formation of nanoparticles.

12.4.1.3 SPUTTERING

Here, nanoparticles are generated by ejecting material from a solid target in a vacuum through the bombardment of energetic particles.

12.4.2 BOTTOM-UP METHODS

These methods assemble nanoparticles from atomic or molecular precursors, allowing for enhanced control over particle size and morphology. Key bottom-up techniques are as follows.

12.4.2.1 CHEMICAL VAPOR DEPOSITION

Volatile precursors decompose or react on a substrate to form nanoparticles.

12.4.2.2 PRECIPITATION

This chemical method involves the reaction of precursors in a solution, resulting in the precipitation of nanoparticles.

12.4.2.3 SOL–GEL PROCESS

Beginning with a chemical solution (sol) that serves as a precursor, nanoparticles form within an evolving network (gel) through various chemical reactions.

12.4.3 BIOLOGICAL METHODS

Biological production involves utilizing microorganisms, plants, or plant extracts to create nanoparticles, a process viewed as more sustainable and environmentally benign.

12.4.3.1 MICROBIAL SYNTHESIS

Microbial synthesis utilizes bacteria, fungi, or algae to convert metal ions into metallic nanoparticles under controlled conditions.

12.4.3.2 PHYTOCHEMICAL SYNTHESIS

Plant extracts, rich in reducing agents such as phenolics, terpenoids, and alkaloids, are employed to reduce metal salts to nanoparticles.

12.5 APPLICATIONS OF NANOPARTICLES IN AQUACULTURE

Nanoparticles serve multiple roles in aquaculture, including production and application that are imperative to fully harness their advantages responsibly. Nanotechnology, renowned for its precision in manipulating materials at the molecular and atomic levels, presents a wide array of opportunities for enhancing freshwater aquaculture. This cutting-edge technology can significantly improve fish health, optimize water quality, and boost feed efficiency, thereby fostering the development of more sustainable and productive aquaculture systems. Here, we delve into several pivotal applications of nanotechnology in freshwater aquaculture (Figure 12.2).

12.5.1 WATER QUALITY MANAGEMENT

The quality of water is a critical factor in aquaculture, directly impacting the health and growth rates of aquatic species. Nanoparticles such as silver and titanium dioxide are utilized for their antimicrobial properties to purify and treat water effectively. Nanotechnology plays a crucial role in this domain as mentioned below.

12.5.1.1 NANOFILTERS

These filters, operating at the nanoscale, are more effective than traditional filters at eliminating pollutants, heavy metals, and pathogens, and they aid in water recycling efforts to promote a more sustainable aquaculture environment.

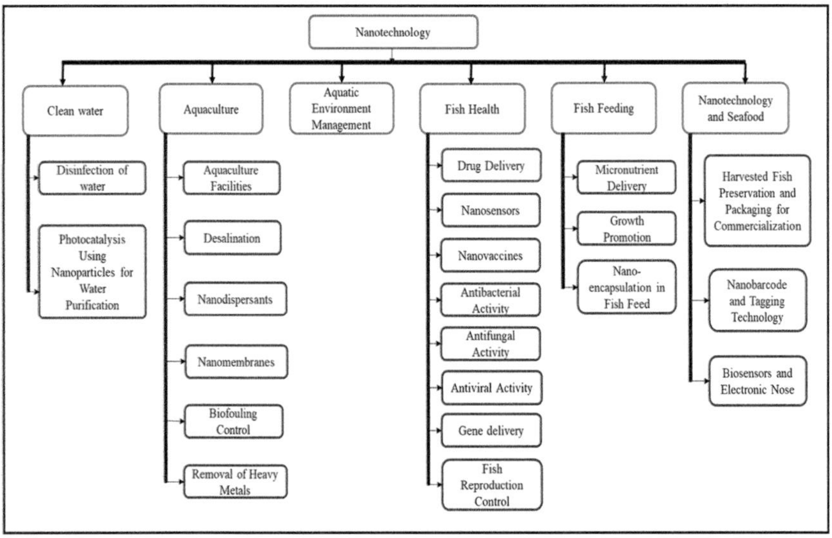

FIGURE 12.2 Applications of nanotechnology in aquaculture.

12.5.1.2 NANOSENSORS

These instruments are capable of detecting contaminants and fluctuations in water parameters (such as pH, dissolved oxygen, and toxic substances) at extremely low concentrations. The ability to monitor these parameters in real time allows for immediate adjustments to maintain optimal conditions.

12.5.2 DISEASE MANAGEMENT

Outbreaks of disease can have devastating effects on aquaculture operations. Nanoparticles facilitate targeted drug and vaccine delivery to specific organs or tissues, enhancing treatment efficacy and minimizing side effects (De Jong and Borm, 2008). Nanotechnology offers the following advanced solutions for both preventing and treating diseases.

12.5.2.1 NANOPARTICLE-BASED VACCINES

These vaccines are potentially more effective because their small size allows them to penetrate cells more easily and elicit stronger immune responses.

12.5.2.2 NANOSCALE DRUG DELIVERY SYSTEMS

These systems enable the precise delivery of therapeutic agents directly to targeted sites within the fish, enhancing treatment effectiveness while reducing dosages and adverse effects.

12.5.3 FEED ENHANCEMENT

Efficient utilization of feed is essential for minimizing costs and environmental impacts associated with aquaculture (Bhoopathy et al., 2021). Nanoparticles are incorporated into feeds to enhance nutrient absorption and feed efficiency, with nanoencapsulation techniques increasing the bioavailability of essential nutrients.

12.5.3.1 NANOPARTICLES IN FEED

Incorporating nanoparticles into fish feed can improve the nutritional value, enhance feed conversion ratios, and promote faster and healthier growth.

12.5.3.2 NANOSCALE ADDITIVES

Nutrients critical for fish health, such as vitamins and minerals, can be encapsulated in nanoparticles to increase their stability and bioavailability.

12.5.4 BREEDING AND GENETICS

Nanotechnology also facilitates genetic enhancements in fish stocks.

12.5.4.1 GENE DELIVERY SYSTEMS

Employing nanotechnology can lead to the development of more efficient and less invasive gene delivery techniques, supporting selective breeding initiatives aimed at improving traits such as growth rate, disease resistance, and stress tolerance.

12.5.5 AQUACULTURE EQUIPMENT ENHANCEMENT

Nanotechnology can be used to improve the materials used in aquaculture infrastructure.

12.5.5.1 ANTIFOULING COATINGS

Applied to nets and tanks, these coatings prevent biofouling, thus maintaining water flow and quality.

12.5.5.2 ENHANCED DURABILITY

Nanocomposite materials can be utilized to construct more durable aquaculture equipment that withstands harsh environmental conditions, enhancing the longevity and reliability of aquaculture apparatus.

12.5.6 NANOVACCINES AS A NEW TECHNOLOGY IN THE FIELD OF FISH HEALTH MANAGEMENT

The convergence of nanotechnology and biotechnology has significantly advanced biomedicine and led to the creation of nanovaccinology, a field that develops nanovaccines. These vaccines incorporate antigens within nanoparticles to enhance immune system targeting and control infection spread. Traditional vaccines used whole pathogens, which caused unwanted side effects. Modern nanovaccines use only essential antigens, reducing side effects but previously lowering immunogenicity (Baltic et al., 2013). Now, nanoparticles are used as carriers or adjuvants to enhance the immune response due to their similar size to pathogens. Nanovaccines improve stability in blood, extend shelf life, eliminate the need for booster doses, and offer targeted delivery, making them more effective for disease prevention.

In aquaculture, nanoparticle-based vaccines provide multiple advantages for vaccinating farmed fish, promising enhanced protection against diseases. However, their environmental and health impacts warrant caution. Vaccination remains critical in aquaculture for managing infectious diseases, despite difficulties in developing effective vaccines for many pathogens. The industry's reliance on antibiotics to combat diseases in salmon farming has raised sustainability concerns.

12.5.7 NANOSELENIUM PARTICLES IN THE FIELD OF AQUACULTURE

Research across various fish species, including crucian carp, artemia, and rainbow trout, has shown that selenium nanoparticles (SeNPs) offer significant nutritional benefits over organic and inorganic selenium forms (Bhoopathy et al., 2021). Studies indicate that SeNPs enhance growth efficiency, muscle content, and antioxidant mechanisms more effectively. They have been found to improve reproduction, digestion, development, and immune responses, while also reducing oxidative stress and increasing the overall productivity in both fisheries and livestock.

Dietary supplementation with SeNPs at 0.2 mg/kg has been shown to enhance weight gain and improve feed efficiency while bolstering resistance to hypoxia and boosting immunity. Additionally, SeNPs have been demonstrated to increase thermal tolerance and enhance metabolic enzyme activity under environmental stress such as exposure to high temperatures and lead. While moderate levels of SeNPs (~4 mg/kg dry matter) can improve the rearing efficiency of fish larvae, higher doses may have detrimental effects. Overall, SeNPs present a potent tool for enhancing aquaculture productivity and resilience.

12.5.8 NANOTECHNOLOGY DEVICES FOR AQUATIC ENVIRONMENT MANAGEMENT

Nanotechnology applications in seawater shrimp aquaculture have proven to be effective in enhancing water quality and shrimp survival, significantly reducing water exchange rates (Dawood et al., 2021). Nanonet treatments notably doubled shrimp survival rates and substantially reduced water nitrite and nitrate levels, while also increasing water pH and efficiency. In China, Nano-863, a high-tech agricultural nanoproduct, is widely used across livestock, crop cultivation, and aquaculture sectors.

12.5.9 NANOTECHNOLOGY AS A NEW INVENTION IN FISH DISEASE TREATMENT

Nanotechnology, which manipulates and produces materials on a nanometer scale—a billionth of a meter—has vast applications across textiles, electronics, engineering, and medicine due to its unique physical and chemical properties derived from size, shape, and surface characteristics (Can et al., 2011).

In aquaculture, nanotechnology plays a crucial role in disease prevention and monitoring. It is used for creating antibacterial and antifungal surfaces, detecting pathogens with nanosensors, and enhancing the delivery of veterinary products through nanotechnology-enhanced fish feeds. Notably, nanotrace elements in feeds are absorbed more efficiently than traditional inorganic trace elements, often doubling the usage level. Furthermore, nanomaterials like chitosan are used for nanoencapsulating vaccines, offering effective protection and delivery against aquatic pathogens. These technologies are pivotal in cultivating pathogen-free aquatic organisms, significantly advancing sustainable aquaculture practices.

12.5.10 ENHANCEMENT OF FISH GROWTH INDICES

Nanotechnology holds significant promise in aquaculture and seafood industries, enhancing growth and health in marine species. Research has demonstrated that young carp and sturgeon fed with iron nanoparticles exhibit accelerated growth rates. Similarly, nanoselenium supplementation in diets has improved weight, growth rate, and antioxidant levels in crucian carp, surpassing results obtained with traditional selenium sources like selenomethionine. Additionally, while direct application of silver nanoparticles in water has proven to be detrimental to young trout, using silver nanoparticle-coated filters has effectively prevented fungal infections in rainbow trout. Beyond disease management, nanotechnology applications extend to enhancing fish food delivery systems, equipping aquaculture with antifouling nets, and developing antibacterial agents for tanks (Feng et al., 2010). These advancements also include innovative packaging materials for seafood shipping and equipment for monitoring seafood freshness, demonstrating nanotechnology's broad utility in improving food quality and operational efficiency in aquaculture settings.

12.5.11 WATER FILTRATION, PURIFICATION, AND REMEDIATION

Nanoenabled technologies leverage nanomaterials like activated carbon, alumina, zeolite, and iron compounds to remove contaminants such as ammonia, nitrites, and nitrates from aquaculture water (Crab et al., 2009). Ultrafine iron powder is also employed to detoxify harmful organic compounds, thereby advancing nanoaquaculture practices.

12.5.12 HARVESTING OF FISH

Traditional fishing lures, enhanced by a polyimide nanocoating after coloring, reflect light multidirectionally, significantly increasing fish catch probability by two to three times.

12.5.13 BIOFOULING CONTROL

Nanotechnology enhances aquaculture by improving disease control, feed formulation, and biofouling control. Coatings with metal-oxide nanoparticles such as ZnO, CuO, and SiO_2 create effective antifouling surfaces, used in fishing nets, aquaculture tank treatments, and marine product packaging.

12.5.14 REMOVAL OF HEAVY METALS CONTAMINATION

Ligand-based nanocoatings, utilizing bifunctional self-assembling ligands, are effectively used for heavy metal removal from water due to their high absorption capacity and cost-effectiveness (Can et al., 2011). These coatings can be regenerated *in situ*, making them sustainable options for water purification. With their extensive surface area and reactive properties, nanomaterials like metal-oxide nanoparticles—including ferric, aluminum, manganese, cerium, magnesium, and titanium oxides—are widely researched for their efficiency in adsorbing heavy metals from aqueous systems. The development of these nanoparticles and their integration into composite materials or granular forms continue to be a focus in scientific research, driven by the need to understand their adsorption mechanisms through advanced analytical techniques such as X-ray absorption spectroscopy and numerical manifold method.

Additionally, the interaction between humic and fulvic acids and various nanomaterials, especially iron-based, carbon-based, and photo-catalytic nanomaterials, has been explored to enhance heavy metal removal from water, highlighting the complex environmental interactions. Chitosan nanoparticles, particularly when combined with clays such as kaolinite, bentonite, and montmorillonite or integrated into magnetite nanocomposites, have also shown promising results in heavy metal adsorption, offering potential improvements in water treatment technologies.

12.6 CHALLENGES AND CONSIDERATIONS

Despite their potential, the use of nanoparticles in aquaculture faces challenges related to environmental impact, toxicity, and public acceptance. Developing regulatory frameworks and conducting comprehensive risk assessments are crucial to ensure the safe application of nanoparticles in aquaculture ecosystems. While nanotechnology promises transformative advancements for freshwater aquaculture, its integration comes with significant challenges and ethical considerations that are mentioned below.

12.6.1 REGULATION AND SAFETY

Introducing nanomaterials into aquatic environments demands stringent regulatory oversight to prevent potential toxicity to aquatic life and ecosystems.

12.6.2 COST

The research, development, and implementation of nanotechnologies can be costly, and these financial expenditures must be justified by the benefits they offer.

12.6.3 PUBLIC PERCEPTION

There are potential concerns from consumers about the use of nanotechnology in food production systems. Addressing these concerns requires transparent communication and comprehensive safety evaluations.

12.7 A WAY FORWARD

The future of nanotechnology in aquaculture holds promise for further innovations in improving aquaculture sustainability and productivity. Researchers continue to explore new nanomaterials and their applications, striving to optimize their effectiveness while ensuring safety and minimizing environmental impacts. The ongoing development and discovery of nanotechnology applications in aquaculture reflect a broader trend toward more technologically advanced, efficient, and sustainable farming practices. As the field grows, it will likely continue to unveil new tools and methods that could significantly transform the aquaculture industry.

12.8 CONCLUSION

Nanotechnology offers powerful tools that have the potential to revolutionize freshwater aquaculture, enhancing sustainability, efficiency, and productivity. However, its successful adoption is contingent upon rigorous research, robust regulatory frameworks and effective public engagement to ensure its benefits are realized safely and responsibly. Overall, nanotechnology holds significant promise for transforming aquaculture into a more productive and sustainable industry, but its application must be carefully managed to mitigate potential risks.

KEYWORDS

- **nanotechnology**
- **freshwater aquaculture**
- **nanovaccines**
- **nanoparticles**
- **water quality management**

REFERENCES

Alexander S, Mewhinney M (2008) NASA Nanotechnology-Based Biosensor Helps Detect Biohazards. RELEASE: 08–131

Baltic MZ, Boskovic M, Ivanovic J et al (2013) Nanotechnology and its potential applications in meat industry. *Tehnologija Mesa* 54:168–175

Bhoopathy, S., Inbakandan, D., Thirugnanasambandam, R., Kumar, C., Sampath, P., Bethunaickan, R., Raguraman, V. and Vijayakumar, G.K., 2021. A comparative study on chitosan nanoparticle synthesis methodologies for application in aquaculture through toxicity studies. *IET Nanobiotechnology*, *15*(4), pp.418–426.

Branton D, Deamer DW, Marziali A et al (2008) The potential and challenges of nanopore sequencing. *Nature Biotechnology* 26:1146–1153

Can, E., Kizak, V., Kayim, M., Can, S.S., Kutlu, B., Ates, M., Kocabas, M. and Demirtas, N., 2011. Nanotechnological applications in aquaculture-seafood industries and adverse effects of nanoparticles on environment. *Journal of Materials Science and Engineering*, *5*(5).

Crab R, Kochva M, Verstraete W, Avnimelech Y (2009b) Bio-flocs technology application in over-wintering of tilapia. *Aquacultural Engineering* 40:105–112

Dawood, M.A., Basuini, M.F.E., Yilmaz, S., Abdel-Latif, H.M., Kari, Z.A., Abdul Razab, M.K.A., Ahmed, H.A., Alagawany, M. and Gewaily, M.S., 2021. Selenium nanoparticles as a natural antioxidant and metabolic regulator in aquaculture: a review. *Antioxidants*, *10*(9), p.1364.

De Jong WH, Borm PJ (2008) Drug delivery and nanoparticles: applications and hazards. *International Journal of Nanomedicine* 3(2):133–149

Deng Y, Cheng Q (2003) Affects of nano-selenium on the growth of Nile tilapia (Oreochromis niloticus). *Inland Aquat Prodd* 6:28–30

Elechiguerra JL, Burt JL, Morones JR et al (2005) Interaction of silver nanoparticles with HIV-*1. Journal of Nanobiotechnology* 1:3–6

Feng C, Khulbe KC, Matsuura T (2010) Recent progress in the preparation, characterization, and applications of nanofibers and nanofiber membranes via electrospinning/interfacial polymerization. *Journal of Applied Polymer Science* 115:756–776

Gunalan S, Sivaraj R, Rajendran V (2012) Green synthesized ZnO nanoparticles against bacterial and fungal pathogens. *Progress in Natural Science: Materials International* 22(6):693–700

Khosravi-Katuli, K., Prato, E., Lofrano, G., Guida, M., Vale, G. and Libralato, G., 2017. Effects of nanoparticles in species of aquaculture interest. *Environmental Science and Pollution Research*, *24*, pp.17326–17346.

Márquez, J.C.M., Partida, A.H., del Carmen, M., Dosta, M., Mejía, J.C. and Martínez, J.A.B., 2018. Silver nanoparticles applications (AgNPS) in aquaculture. *International Journal of Fisheries and Aquatic Studies*, *6*(2), pp.5–11.

Nasr-Eldahan, S., Nabil-Adam, A., Shreadah, M.A., Maher, A.M. and El-Sayed Ali, T., 2021. A review article on nanotechnology in aquaculture sustainability as a novel tool in fish disease control. *Aquaculture International*, *29*, pp.1459–1480.

Noor, N.N.M., Kamaruzaman, N.H., Al-Gheethi, A., Mohamed, R.M.S.R. and Hossain, M.S., 2023. Degradation of antibiotics in aquaculture wastewater by bio-nanoparticles: A critical review. *Ain Shams Engineering Journal*, *14*(7), p.101981.

Ogunfowora, L.A., Iwuozor, K.O., Ighalo, J.O. and Igwegbe, C.A., 2021. Trends in the treatment of aquaculture effluents using nanotechnology. *Cleaner Materials*, *2*, p.100024.

Samanta, P., Dey, S., Ghosh, A.R. and Kim, J.K., 2022. Nanoparticle nutraceuticals in aquaculture: A recent advances. *Aquaculture*, *560*, p.738494.

Sarkar, B., Mahanty, A., Gupta, S.K., Choudhury, A.R., Daware, A. and Bhattacharjee, S., 2022. Nanotechnology: A next-generation tool for sustainable aquaculture. *Aquaculture*, *546*, p.737330.

Selvaraj, B., Subramanian, K., Gopal, S. and Renuga, P.S., 2014. Nanotechnology as a novel tool for aquaculture industry: a review. *World Journal of Pharmaceutical Sciences*, pp.1089–1096.

Shah, B.R. and Mraz, J., 2020. Advances in nanotechnology for sustainable aquaculture and fisheries. *Reviews in Aquaculture*, *12*(2), pp.925–942.

Tayel, A.A., Elsaied, B.E. and Diab, A.M., 2019. Nanotechnology for aquaculture. *Nanoscience for Sustainable Agriculture*, pp.479–544.

CHAPTER 13

Interactive Effects of Warming and Pollution on Aquatic Environment

NEHA GANGAN[1*], EDIGA ARUN GOUD,[2] and BHAGCHAND CHHABA[3]

[1]*Department of Aquatic Environment Management, Dr. Balasaheb Sawant Konkan Krishi Vidyapeeth, Dapoli, College of Fisheries, Shirgaon, Ratnagiri, Maharashtra, India*

[2]*College of Fisheries, Dr. Balasaheb Sawant Konkan Krishi Vidyapeeth, Ratnagiri, Maharashtra, India*

[3]*LSPN College of Fisheries, Kawardha, Chhattisgarh, India*

*Corresponding author

ABSTRACT

Amid the growing environmental challenges, urbanization and industrialization are the major contributors to warming and pollution, independently impacting aquatic environments significantly. However, their interactive effects proved to be more substantial, increasing ecological vulnerabilities and impacting the hydrological cycle, temperature dynamics, biogeochemical cycles, and aquatic biodiversity. Interactive impacts include altered hydrological regimes, physiological and behavioral changes, increased harmful algal blooms, and disruptions in food web dynamics. Thus, comprehensive research and proactive measures are crucial to comprehend and mitigate the synergistic impacts of warming and pollution, ensuring the resilience and sustainability of aquatic environments.

Advances in Freshwater Aquaculture: Technology for Sustainable Development. Tarang Kumar Shah, Madhu Sharma, Mohd. Ashraf Rather, & Avdhesh Kumar (Eds.)

DOI: 10.1201/9781779640857-13

13.1 INTRODUCTION

Water, a unique element of life, sustains all beings on the Earth. Marine environments cover 71%, while freshwater ecosystems cover 1%, together contributing 50% to the global net primary production. Aquatic primary producers, just 1% of the standing crop, play a vital role in biomass, CO_2 assimilation, and oxygen production (Hader et al., 2020). Water's hydrological cycle, transitioning between gas, liquid, and solid states, crucially regulates climate and sustains freshwater. Its high specific heat, seen in oceans acting as thermal reservoirs, moderates temperature fluctuations for a habitable environment. Yet, population growth and industrialization amplify environmental impacts, straining resources, emitting pollutants, and worsening climate change. This, in turn, threatens the delicate balance crucial for diverse life forms (Inglezakis et al., 2016).

Pollution and climate change independently exert significant effects on the aquatic environment. Nevertheless, their synergetic impacts on the aquatic environment are even more substantial (Moe, 2013; Cabral et al., 2019; Gissi et al., 2021). For instance, warmer water temperatures can promote the growth of harmful algal blooms that produce toxins that are harmful to aquatic life and humans. Warming water temperatures can increase the toxicity of pollutants, leading to greater harm to aquatic organisms (Dinh et al., 2022). Overall, the combined effects of pollution and climate change on aquatic environments are complex and can vary depending on the specific ecosystem and stressors involved.

13.2 EFFECTS OF POLLUTION ON AQUATIC ENVIRONMENT

13.2.1 FRESHWATER ENVIRONMENT

Over the past years, freshwater ecosystems have faced significant challenges due to human activities, including industrial discharges, agricultural practices, urban waste management problems, and expanding urbanization, posing serious threats to their health and sustainability (Amoatey and Baawain, 2019). Thus, the pollution of these freshwater bodies has jeopardized the supply of clean drinking water (Hader et al., 2020). Moreover, it has a greater impact on freshwater flora and fauna. Water quality and habitat issues in lakes and streams have the potential to propagate downstream and affect the entire ecosystem, including an influx of excessive nutrients from wastewater discharges, pollutants from stormwater runoff, particularly from roads and urban areas, agricultural runoff, warmer water temperatures due to the loss of riparian areas along streams, and alterations in hydrology or water flow patterns.

Pollution adversely affects the physico-chemical composition of water, which eventually has effects on the aquatic biodiversity. Heavy sewage pollution and effluent discharge with high organic matter into water bodies cause a significant reduction in dissolved oxygen (DO) levels. Deoxygenation and increased turbidity can cause a reduction or elimination of algae. Furthermore, downstream areas of rivers often face significant challenges due to biodegradable organic pollution, particularly affecting migratory fish species like *Salmo salar* and *Salmo trutta*, which require high levels of DO. In some instances, the presence of low DO levels and organic pollutants can lead to avoidance behavior, acting as a barrier that prevents these fish from reaching well-oxygenated breeding and spawning grounds (Malik et al., 2020).

Rapid urbanization and frequent discharge of wastewater into surface water bodies lead to extensive freshwater contamination with emerging organic contaminants (EOCs) such as pharmaceuticals, insecticides, surfactants, and endocrine disruptors including hormones (Pal et al., 2010). Various sources can add to the presence of aqueous EOCs, such as untreated or treated wastewater discharge, runoff from agricultural and livestock operations, industrial processes, urban drainage, and accidental spills. The elimination of EOCs through wastewater treatment remains challenging, expensive, and heavily reliant on specific processes (Richards et al., 2021). Despite their presence at trace levels, the detrimental impacts of these organic contaminants on aquatic life, animals, and humans are becoming an increasingly concerning issue (Pal et al., 2010). Pharmaceuticals and illicit drugs are among the categories of pollutants that affect freshwater organisms. For instance, Tramadol and Citalopram concentration levels alter the behaviors of an aquatic invertebrate like crayfish (Malik et al., 2020).

13.2.2 COASTAL AND MARINE ENVIRONMENT

The Earth's surface is predominantly covered by oceans, accounting for 70% of its area, and they play a vital role in the global environment. Moreover, they are responsible for controlling earth's climate and influencing worldwide biogeochemical processes. The oceans serve various purposes, including serving as places for recreational activities and tourism, acting as transportation routes, as a repository of genetic and biological diversity, and serving as disposal areas for waste materials. The coastal areas of the oceans also partake in these functions (McLean et al., 2001). However, many coastal and marine areas around the world have been documented as being adversely impacted by pollution, which has had a significant negative influence on both commercial fisheries in coastal and marine environments (Islam and Tanaka, 2004).

GESAMP (1973) categorizes studies on marine environmental pollution into impacts on living resources, human health hazards, amenities reduction, and interference with sea users. Natural marine populations undergo annual and extended fluctuations due to factors like offspring survival and cyclic climate-related changes. Therefore, assessing pollution impact on these populations in the open ocean is a challenging task (Cole, 1979).

Understanding pollution impacts on coastal and marine environments necessitates a focus on key pollutants such as fertilizers, pesticides, sewage, persistent organic pollutants, nutrients, oils, radionuclides, heavy metals, pathogens, sediments, and litter (Williams, 1996; Islam and Tanaka, 2004). The global increase in fertilizer use, especially in agriculture, significantly contributes to pollution, leading to nutrient enrichment in coastal waters. Pesticide residues, undergoing bioconcentration and biomagnification in marine trophic chains, result in chronic effects such as organism mortality, cancer, reproductive issues, immune suppression, and damage to cellular and molecular levels, impacting overall fish health (Islam and Tanaka, 2004; Todd et al., 2010).

Marine pollution stems from both land-based factors, including rivers and wind, and activities such as dredging, mining, dumping, and shipping within the sea. Escalating human population, deforestation, and rapid urbanization in various regions result in substantial sediment and pollution loads in coastal waters (Spalding et al., 2001). Sediments from agricultural runoff, deforestation, and construction sites reduce light penetration, impacting photosynthesis in seagrass and corals and affecting predator–prey interactions. Urban runoff, coupled with untreated sewage and industrial discharge, contributes to elevated nitrogen and phosphorus levels, leading to eutrophic conditions (Islam and Tanaka, 2004; Todd et al., 2010).

Various industrial activities release heavy metals and trace elements into the marine ecosystem as byproducts, posing risks to organisms in marine food chains. Heavy metals interfere with cell membrane permeability in phytoplankton and marine algae, compromising cellular integrity (Islam and Tanaka, 2004). Phytoplankton and mangroves, when exposed to heavy metals, experience morphological changes and reduced photosynthetic activity, leading to diminished growth and increased mortality, ultimately impacting the structure and diversity of mangrove ecosystems (Table 13.1) (Kayser, 1976; Bayen, 2012).

TABLE 13.1 Range of Detectable Levels of Contaminants in *Avicennia marina,* Mumbai, India

Pollutant	Reported Range (ng/g)
Hexachlorocyclohexane (HCHs)	4.478–10.475
Dichlorodiphenyltrichloroethane (DDTs)	1.173–15.399
Endosulfans	0.633–1.542
Aldrin	0.177–0.256

In recent years, microplastics (<5 mm) have been accumulating in the world's oceans and sediments. Gyres and the industrial coastal areas have been identified as microplastic hotspots. Being lightweight and durable, plastic has become a pervasive component of the marine debris. Ingestion and entanglement by marine organisms are the prime concerns associated with microplastics. Its consequences include internal abrasions, ulcers, and blockages in the digestive tract, resulting in starvation, satiation, and physical deterioration. These, in turn, can contribute to lowered reproductive fitness, drowning, diminished predator avoidance, impaired feeding ability, and bioaccumulation. Besides this, plastic materials, with their high absorbance efficiency, can accumulate persistent toxic chemicals, posing a risk of transporting these substances into marine organisms through contaminated debris (Wright et al., 2013; Thushari and Senevirathna, 2020).

13.3 EFFECTS OF WARMING ON AQUATIC ENVIRONMENT

13.3.1 FRESHWATER ENVIRONMENT

Warming has a pronounced impact on freshwater ecosystems due to its significant influence on key determinants such as water quality and quantity. Atmospheric temperature regimes directly affect water temperature and various chemical aspects, including DO levels, crucial for supporting freshwater biodiversity and sustaining essential ecological functions and services. Additionally, the warming effect extends to surface and groundwater processes, encompassing precipitation, snowmelt, runoff, soil moisture, river discharge, and aquifer recharge (Capon et al., 2021).

Global warming can induce alterations in water quality and hydrological regimes, including reduced oxygen levels and modified nutrient dynamics, thereby influencing the overall health of freshwater ecosystems; concurrently, shifts in precipitation patterns may affect water quantity, causing changes in river flows and aquifer recharge rates. Altered flow regimes can affect the availability of water for irrigation and impact freshwater fisheries production, posing risks to food security in regions reliant on freshwater resources. Warming and elevated CO_2 levels can result in changes in riparian vegetation such as the encroachment of woody species into nonwooded wetland ecosystem. This can alter the habitat structure and food resources for aquatic organisms, impacting biodiversity and ecosystems functions. Increased water temperatures can exceed the thermal tolerances of coldwater fish species, leading to declines in their populations. Similarly, changes in flow regimes can disrupt the life cycles of aquatic invertebrates, affecting the overall biodiversity of freshwater ecosystems (Capon et al., 2021). In lentic environments elevated air temperatures result in increased surface temperatures, triggering earlier thermal stratification that adversely affects coldwater fish by reducing growth rates, increasing heat mortality, and intensifying food competition among aquatic organisms due to deeper thermoclines (Kernan et al., 2011; Hassan et al., 2020).

13.3.2 COASTAL AND MARINE ENVIRONMENT

Global warming serves as a key stressor within the broader context of climate change, exerting significant impacts on coastal and marine environments by contributing to rising sea levels, ocean acidification, changes in water temperature, and disruptions to ecosystems, ultimately threatening biodiversity, fisheries, and the overall health of marine ecosystems. In addition to serving as the planet's heat reservoir, the oceans have assimilated about a third of the carbon dioxide emitted through human activities. In turn, the upper ocean layers warmed by 0.6°C in the past century, while anthropogenic CO_2 absorption has led to ocean surface acidification. Rising temperatures and ocean acidification can reduce nutrient availability, posing challenges for the survival and reproduction of phytoplankton and primary producers, thereby triggering cascading effects throughout the food web. Elevated temperatures directly affect individuals across different stages of their life cycle by influencing their physiology, morphology, and behavior. Warming's effects also extend to populations through altered transport processes affecting dispersal and recruitment, while community-level impacts involve climate-induced changes in both the abundance and per capita interaction strength of interacting species. For instance, if a predator species moves into a new area where it has no natural predators, it may consume more of its prey than usual, leading to declines in those populations. Moreover, global warming can impact the growth, reproduction, and survival of crucial habitat-forming species, such as coral reefs, seagrass beds, and kelp forests, potentially affecting dependent species relying on these habitats for food, shelter, or breeding sites. These collective impacts lead to emerging ecological responses, encompassing changes in species distributions, biodiversity, productivity, and microevolutionary processes (Harley et al., 2006; Hoegh-guldberg et al., 2010).

In the realm of aquatic ecosystems, regardless of the individual impacts of warming and pollution, their combined effects proved to be more detrimental to the aquatic environments. The interaction of warming and pollution has adverse effects on aquatic environments, disrupting ecosystems, changing the distribution and toxicity of contaminants, and jeopardizing the well-being and survival of diverse species such as fish and corals.

13.4 INTERACTIVE EFFECTS OF WARMING AND POLLUTION

13.4.1 FRESHWATER ENVIRONMENT

Global climate change-induced temperature changes profoundly affect the life-history traits of ectotherms and exert direct or indirect influence on various ecosystem properties, processes, and functions, including DO concentrations, nutrient cycling, and litter decomposition. Furthermore, temperature alterations can impact chemical toxicity by enhancing uptake, metabolism, or excretion rates in organisms, and can also alter fate processes such as degradation rates and adsorption kinetics (Hermann et al., 2023).

The interactive impacts of elevated temperatures and pollution on freshwater ecosystems can have significant and complex effects on various trophic levels and ecosystem processes. They can lead to shifts in community assemblages, affecting the abundance and composition of organisms such as zooplankton (Hermann et al., 2023). However, some studies highlight that existing micropollutants and rising temperatures or heatwaves generally exhibit an antagonistic effect on the growth of primary producers (Polazzo et al., 2022). Another synergistic impact is that increasing temperatures can elevate organisms' susceptibility to pollutants, potentially resulting in higher uptake rates due to enhanced metabolism and increased food consumption. The interactive effects of heatwaves, elevated temperatures, and pollution in freshwater ecosystems give rise to intricate dynamics, manifesting in shifts in organism size and thermal tolerance, modifications in ecosystem processes, and the propagation of significant cause-and-effect chains across various trophic levels (Hermann et al., 2023).

13.4.2 COASTAL AND MARINE ENVIRONMENT

Coastal and marine environments are intricately shaped by the interactive effects of warming and pollution, two interrelated phenomena that pose significant threats to the delicate balance of these ecosystems. Warming can increase the toxicity of pollutants, leading to greater harm to marine organisms. Research findings indicate that the additive effects of warming and eutrophication on the fertilization of the tropical coral *Acropora tenuis* are additive, while the impact on embryo deformities shows a synergistic effect (Cabral et al., 2019). Moreover, the anticipated intensification of organic and inorganic pollution on coral reefs due to climate-driven temperature increases is expected to heighten coral susceptibility to disease and temperature-induced bleaching (Gissi et al., 2021). Concurrently, warming and pollution can have additive or synergistic effects on the growth and survival of marine organisms. Gouvêa et al. (2017) found that the combined effect of eutrophication and warming on macroalgae led to large-scale declines in macroalgal production. A study by Coelho et al. (2013) investigated the interactive effects of anthropogenic pollutants and ocean acidification on a marine microbial community. The results showed that the role that microbes play in biogeochemical cycles might be affected by the interactive effects of these two drivers. Specifically, the study found that the combined effects of pollution and ocean acidification led to changes in the composition and activity of microbial communities, which in turn affected the cycling of nutrients in the environment (Figure 13.1).

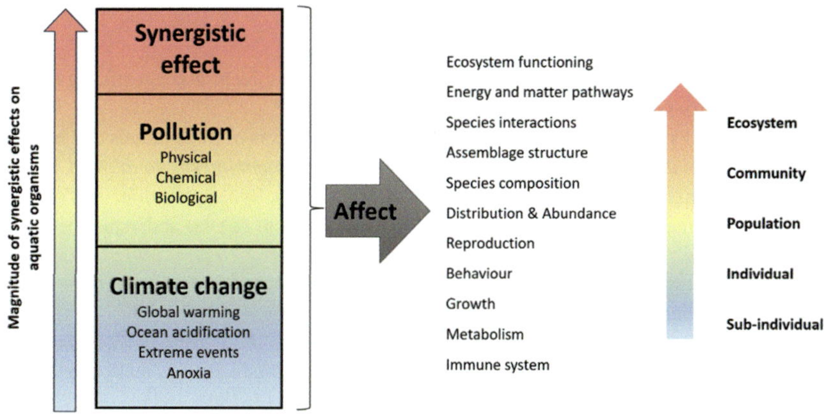

FIGURE 13.1 Synergistic effects of pollution and climate change on aquatic environments.

Source: Reprinted from Cabral et al., 2019. Copyright © 2019 by the authors. Licensee MDPI, Basel, Switzerland. http://creativecommons.org/licenses/by/4.0/

In conclusion, the interactive effects of warming and pollution on aquatic environments underscore the urgency of comprehensive and adaptive conservation measures. As these two stressors synergize to amplify the challenges faced by lakes, rivers, oceans, and other water bodies, the intricate web of ecological disruptions demands a holistic approach. Recognizing the nuanced interplay between warming and pollution is essential for crafting effective strategies that address both stressors concurrently. Only through concerted efforts in research, policy-making, and sustainable practices can we hope to mitigate the cumulative impacts and ensure the long-term health and resilience of our precious aquatic environments in the face of a rapidly changing environment.

KEYWORDS

- **aquatic ecosystem**
- **bioaccumulation**
- **pollution and warming**

REFERENCES

Amoatey, P., & Baawain, M. S. (2019). Effects of pollution on freshwater aquatic organisms. *Water Environment Research*, *91*(10), 1272–1287.

Bayen, S. (2012). Occurrence, bioavailability and toxic effects of trace metals and organic contaminants in mangrove ecosystems: a review. *Environment International*, *48*, 84–101.

Cabral, H., Fonseca, V., Sousa, T., & Costa Leal, M . (2019). Synergistic effects of climate change and marine pollution: An overlooked interaction in coastal and estuarine areas. *International Journal of Environmental Research and Public Health*, *16*(15), 2737.

Capon, S. J., Stewart-Koster, B., & Bunn, S. E. (2021). Future of freshwater ecosystems in a 1.5 C warmer world. *Frontiers in Environmental Science*, *9*, 596.

Coelho, F. J., Santos, A. L., Coimbra, J., Almeida, A., Cunha, A., Cleary, D. F., Calado, R., & Gomes, N. C. (2013). Interactive effects of global climate change and pollution on marine microbes: The way ahead. *Ecology and Evolution*, *3*, 1808–1818.

Cole, H. A . (1979). Pollution of the sea and its effects. *Proceedings of the Royal Society of London. Series B. Biological Sciences*, *205*(1158), 17–30.

Dinh, K. V., Konestabo, H. S., Borgå, K., Hylland, K., Macaulay, S. J., Jackson, M. C., Verheyen, J., & Stoks, R. (2022). Interactive effects of warming and pollutants on marine and freshwater invertebrates. *Current Pollution Reports*, *8*(4), 341–359.

Gesamp. (1973). Report of the Fifth Session. *GESAMP*, *V/10*, 87.

Gissi, E., Manea, E., Mazaris, A. D., Fraschetti, S., Almpanidou, V., Bevilacqua, S., Coll, M., Guarnieri, G., Lloret-Lloret, E., Pascual, M., & Petza, D. (2021). A review of the combined effects of climate change and other local human stressors on the marine environment. *Science of the Total Environment*, *755*, 142–564.

Gouvêa, L. P., Schubert, N., Martins, C. D. L., Sissini, M., Ramlov, F., Rodrigues, E. R. d. O., Bastos, E. O., Freire, V. C., Maraschin, M., Carlos Simonassi, J., et al . (2017). Interactive effects of marine heatwaves and eutrophication on the ecophysiology of a widespread and ecologically important macroalga. *Limnology and Oceanography*, *62*, 2056–2075.

Häder, D. P., Banaszak, A. T., Villafañe, V. E., Narvarte, M. A., González, R. A., & Helbling, E. W. (2020). Anthropogenic pollution of aquatic ecosystems: Emerging problems with global implications. *Science of the Total Environment*, *713*, 136–586.

Harley, C. D., Randall Hughes, A., Hultgren, K. M., Miner, B. G., Sorte, C. J., Thornber, C. S., Rodriguez, L. F., Tomanek, L., & Williams, S. L. (2006). The impacts of climate change in coastal marine systems. *Ecology Letters*, *9*(2), 228–241.

Hassan, B., Qadri, H., Ali, M. N., Khan, N. A., & Yatoo, A. M. (2020). Impact of climate change on freshwater ecosystem and its sustainable management. *Fresh Water Pollution Dynamics and Remediation*, 105–121.

Hermann, M., Peeters, E. T., & Van den Brink, P. J. (2023). Heatwaves, elevated temperatures, and a pesticide cause interactive effects on multi-trophic levels of a freshwater ecosystem. *Environmental Pollution*, *327*, 121–498.

Hoegh-Guldberg, O., & Bruno, J. F. (2010). The impact of climate change on the world's marine ecosystems. *Science*, *328*(5985), 1523–1528.

Inglezakis, V. J., Poulopoulos, S. G., Arkhangelsky, E., Zorpas, A. A., & Menegaki, A. N. (2016). Aquatic environment. *Environment and Development*, *137*–212.

Islam, M. S., & Tanaka, M. (2004). Impacts of pollution on coastal and marine ecosystems including coastal and marine fisheries and approach for management: A review and synthesis. *Marine Pollution Bulletin*, *48*(7–8), 624–649.

Kayser, H. (1976). Waste-water assay with continuous algal cultures: the effect of mercuric acetate on the growth of some marine dinoflagellates. *Marine Biology*, *36*, 61–72.

Kernan, M., Battarbee, R. W., & Moss, B. R. (Eds.). (2011). *Climate Change Impacts on Freshwater Ecosystems*. John Wiley & Sons.

Malik, D. S., Sharma, A. K., Sharma, A. K., Thakur, R., & Sharma, M. (2020). A review on impact of water pollution on freshwater fish species and their aquatic environment. *Advances in Environmental Pollution Management: Wastewater Impacts and Treatment Technologies*, *1*, 10–28.

McLean, R. F., Tsyban, A., Burkett, V., Codignotto, J. O., Forbes, D. L., Mimura, N., Beamish, R. J., & Ittekkot, V. (2001). Coastal zones and marine ecosystems. *Climate Change*, 343–379.

Moe, S. J., De Schamphelaere, K., Clements, W. H., Sorensen, M. T., Van den Brink, P. J., & Liess, M. (2013). Combined and interactive effects of global climate change and toxicants on populations and communities. *Environmental Toxicology and Chemistry*, *32*(1), 49–61.

Pal, A., Gin, K. Y. H., Lin, A. Y. C., & Reinhard, M. (2010). Impacts of emerging organic contaminants on freshwater resources: review of recent occurrences, sources, fate and effects. *Science of the Total Environment*, *408*(24), 6062–6069.

Polazzo, F., Roth, S. K., Hermann, M., Mangold-Döring, A., Rico, A., Sobek, A., Van den Brink, P. J., & Jackson, M. C. (2022). Combined effects of heatwaves and micropollutants on freshwater ecosystems: Towards an integrated assessment of extreme events in multiple stressors research. *Global Change Biology*, *28*(4), 1248–1267.

Richards, L. A., Kumari, R., White, D., Parashar, N., Kumar, A., Ghosh, A., Kumar, S., Chakravorty, B., Lu, C., Civil, W., & Lapworth, D. J. (2021). Emerging organic contaminants in groundwater under a rapidly developing city (Patna) in northern India dominated by high concentrations of lifestyle chemicals. *Environmental Pollution*, *268*, 115–765.

Spalding, M. D., Ravilious, C., & Green, E. P. (2001). *World Atlas of Coral Reefs*. UNEP-WCMC and the University of California Press: Berkley.

Thushari, G. G. N., & Senevirathna, J. D. M. (2020). Plastic pollution in the marine environment. *Heliyon*, *6*(8).

Todd, P. A., Ong, X., & Chou, L. M. (2010). Impacts of pollution on marine life in Southeast Asia. *Biodiversity and Conservation*, *19*, 1063–1082.

Williams, C. (1996). Combating marine pollution from land-based activities: Australian initiatives. *Ocean & Coastal Management* , *33* (1–3), 87–112.

Wright, S. L., Thompson, R. C., & Galloway, T. S. (2013). The physical impacts of microplastics on marine organisms: A review. *Environmental Pollution*, *178*, 483–492.

CHAPTER 14

Integrated Water Resources Management: A Blueprint for Water Resource Sustainability in Aquaculture

VIVEK R. TANDEL[1*], D. T. VAGHELA[1], RITIKA A. TANDEL[2], and BHAVIKA R. TANDEL[3]

[1]*Department of Aquatic Environment Management, College of Fisheries Science, Kamdhenu University, Veraval, Gujarat, India*

[2]*Department of Fish Processing Technology, College of Fisheries Science, Kamdhenu University, Veraval, Gujarat, India*

[3]*Department of Fisheries Resource Management, College of Fisheries Science, Kamdhenu University, Veraval, Gujarat, India*

*Corresponding author

ABSTRACT

Water shortage is a phenomenon that has been observed in recent trends, characterized by a significant decline in water supply and substantial water losses resulting from ineffective management and contamination. The livelihoods of individuals worldwide have been significantly impacted by this issue. Integrated water resources management (IWRM) has emerged as one of the most widely accepted techniques at international water conferences and forums dedicated to exploring options for sustainable water management. IWRM promotes a comprehensive and inclusive strategy that includes users, planners, and policymakers at all levels. The fundamental goals of IWRM include fair access, sustainable water usage

Advances in Freshwater Aquaculture: Technology for Sustainable Development. Tarang Kumar Shah, Madhu Sharma, Mohd. Ashraf Rather, & Avdhesh Kumar (Eds.)

DOI: 10.1201/9781779640857-14

in aquaculture, and environmental conservation. It runs on guiding ideas including adaptive management, resource integration, and stakeholder participation. Data gathering, stakeholder involvement, scenario analysis, and policy development are just a few of the significant phases and duties involved in developing an IWRM plan. The IWRM approach places a strong emphasis on the value of comprehensive planning that takes into account the social, economic, and ecological aspects of water resources in aquaculture.

14.1 INTRODUCTION

Water is a vital component for sustaining life for all living beings. Even though water is abundant on Earth, with over 70% of the planet's surface covered by it, the amount of fresh water available to support human society is relatively small. Only 2.53% of the total water on Earth is freshwater, and out of this, 1.74% is frozen in the Antarctic and Arctic ice sheets, as well as in mountain glaciers (Sun et al., 2011). Despite fresh water being considered a renewable resource due to the natural water cycle, the world faces an imminent water crisis due to several factors. These factors include the changing global environment, the continuous growth of the world's population, and the severe environmental pollution. As a result, the availability and quality of fresh water are becoming increasingly uncertain and challenging to manage. The concept of integrated water resources management (IWRM) originated during the *United Nations Conference on Water* in 1977, held in Mar Del Plata, Argentina (Rahi and Jazi, 2021).

The last century has seen an increase in water demand that is more than twice as fast as population growth. Although there has not been a global water crisis, over 40% of the world's population—2.8 billion people—live in river basins where there are varied degrees of water shortages. With the withdrawal of more than 75% of river flows, approximately 1.2 billion people face physical water scarcity (Kasbohm et al., 2009)

The Global Water Partnership (GWP) provided the widely accepted definition of IWRM, describing it as a process that fosters the integrated development and management of water, land, and related resources. The ultimate goal is to optimize economic and social welfare while ensuring equity and safeguarding the sustainability of crucial ecosystems (Biswas, 2004). IWRM is a structured approach that ensures the allocation, sustainable development, and monitoring of water resources while considering social, economic, and environmental objectives. IWRM has emerged as the primary framework for water governance, influencing water policies at both national and international levels (Gerlak and Mukhtarov, 2015). Currently, IWRM is integrated into international political agendas and recognized as a crucial element in achieving the sustainable development goals.

14.2 IWRM PRINCIPLES

Water and environmental management integration requires a comprehensive systems approach with active participation from all parties involved, including local residents and employees. Setting social elements as a top priority is essential, with a focus on developing capacity and anticipating trends by having access to pertinent information. Establishing and maintaining an enabling environment requires central government support in addition to the implementation of full-cost pricing augmented by targeted subsidies (Table 14.1). Adopting the finest techniques and technologies available is essential, and it needs to be backed by stable and ongoing funding (Meran et al., 2021). Recognizing water as an economic benefit means that water supplies must be distributed fairly. Furthermore, a thorough and successful strategy requires the empowerment of women in the water management sector.

14.3 IWRM AIM

1. The primary objectives of IWRM are to ensure the sustainability of water resources for environmental purposes, promote equitable access to water for social reasons, and enhance water use efficiency for economic benefits.
2. IWRM seeks to achieve a balance between meeting the increasing water demands of a growing population while simultaneously safeguarding and preserving water resources to ensure a sustainable water supply (Godinez-Madrigal et al., 2019; Hooper, 2005).

The three "Es" of IWRM—equity, ecological integrity, and efficiency—highlight important facets of appropriately managing water resources.

Equity recognizes water as a basic need and emphasizes everyone's fundamental right to access it in adequate quantity and quality. Ecological integrity emphasizes sustainable activities for the benefit of both the current and future generations, with a focus on preserving appropriate water supplies of acceptable quality in the environment. Efficiency, on the other hand, recognizes that water is a limited resource and encourages wise use of it. To encourage responsible and effective management of water resources, it is necessary to achieve cost recovery in water services and match water pricing with its economic value (Meran et al., 2021).

TABLE 14.1 Basic Principles of IWRM

Principle	Description
1	IWRM is implemented on a river basin scale for effective management.
2	IWRM employs a systemic approach, acknowledging the interconnected components that influence each other.
3	Participation involves all users and stakeholders in water resource management.
4	The social dimension is integral to IWRM's framework.
5	Constructive capacity addresses the knowledge and skill gaps hindering IWRM implementation.
6	Adequate information availability and utilization capability are essential.
7	Reliable funding, cost recovery, and subsidies promote solidarity and equity in IWRM.
8	Equitable resource allocation is a key principle.
9	Water is recognized as an economic good in IWRM.
10	Empowering women's roles is emphasized in water management.
11	Government support ensures adaptable and flexible institutional structures.
12	A trans- and multidisciplinary approach fosters collaboration across diverse sciences.
13	IWRM strives for a balance among social, economic, and environmental values.

Source: Marquez-Romance et al. (2022)

14.4 IWRM PROCESS

The application of IWRM can be divided into three major sequential stages (Rahi and Jazi, 2021):

1. Planning: During the planning stage, water resource issues are assessed, national goals for water management are set, and appropriate policies and strategies are developed.
2. Implementation: It focuses on executing the proposed management actions, while also providing capacity-building activities to ensure expertise in using necessary tools and methodologies.
3. Evaluation and monitoring: It includes developing indicators to assess the progress of implemented actions and activities.

14.5 IWRM PARADIGM

The paradigm of IWRM incorporates important concepts including participation, decentralization, integration, and sustainability. IWRM highlights the necessity for integrated management across horizontal sectors, such as water supply, sanitation, agriculture, energy generation, industrial usage, and environmental protection, because of its holistic viewpoint and how these sectors either use or impact water resources. To coordinate efforts among local, regional, national, and international water user organizations and institutions, vertical integration is also called for in addition to horizontal integration. In addition to the necessity of integration, it is believed that increasing awareness of local and regional issues requires decentralized decision-making and accountability at the highest level of efficient management.

As such, IWRM aims to achieve a balance between top-down and bottom-up management, with a focus on strengthening associations of water users and community-based organizations. One of the main tenets of IWRM is sustainability, which goes beyond ecological issues to include financial and economic sustainability. This suggests that the economic value of water must be taken into consideration when allocating resources, requiring the pricing of water to reflect its entire cost (Xie, 2006).

14.6 IWRM GENERAL FRAMEWORK (GWP-TAC, 2000)

In order to successfully apply the IWRM paradigm in real-world settings, the GWP recommended in 2004 that an IWRM planning cycle be followed.

To put it simply, the complexity of the water cycle and the way the water industry is linked to other industries, such as food and energy, calls for particular methods to integrate social, environmental, and economic factors at the watershed level. IWRM provides essential multidisciplinary instruments from engineering, social sciences like political science, sociology, and economics, and natural water sciences like hydrology, geohydrology, and meteorology. Many of these approaches, like decision support systems or optimization models, depend on mathematical models as a necessary precondition to understand the intrinsic complexity of the interdependencies involved (GWP, 2004).

Hydro-economic models rooted in mathematics, functioning as integral tools within the framework of IWRM, frequently utilize simulation or optimization models alongside node-link networks. These models aim to emulate the spatial arrangement of crucial elements within a system, encompassing natural water bodies like seas, lakes, aquifers, and river sections, as well as artificial water bodies such as canals. Additionally, the models incorporate infrastructure components like wells, dams, pipelines, pumps, and purification plants, along with accounting for human-made or artificial impacts on the water system, including points of use, pollution sources (both point and nonpoint), among other factors (Meran et al., 2021) (Figure 14.1).

14.7 IWRM IMPLEMENTATION

IWRM deployment requires a number of crucial components in order to be successful. Develop well-defined basin management plans and have a clear strategy for how to coordinate the development of water resources across sectors and stakeholders. Involving underrepresented stakeholders through consensus-building strategies is also essential, as is capacity building and training, especially at the local government level. The environmental and human elements of the basin ought to be thoroughly understood by scientists. It is important to build mechanisms for coordination and involvement, such as adaptive engagement and stakeholder identification. For effective water resource management, which aims to promote water safety and improved conditions in the drainage region, continuous monitoring, evaluation, and the application of cutting-edge technology are crucial.

FIGURE 14.1 Framework of IWRM.

14.7.1 IMPORTANT ELEMENTS FOR THE IWRM IMPLEMENTATION PROCESS (ANUKULARMPHAI, 2010)

- Unambiguous and shared understanding of IWRM
- Promote awareness and engage in collaborative processes to generate outcomes
- Embed the three fundamental principles of IWRM
- Institutionalize IWRM within the government system through top-level endorsement
- Appoint water "champions" to drive and advance the policy and implementation procedures

14.7.2 PROCESS OF DEVELOPING AN IWRM STRATEGY (Sun et al., 2011)

Steps	Tasks
Determine current standing and articulate overarching objectives	Water resource challenges and advancements in achieving IWRM objectives
Cultivate dedication to overhaul processes	Government determination, public consciousness, and inclusive discussions involving various stakeholders
Analyze gaps	Water resource management functions required in aquaculture
Develop a strategic framework and implement an actionable plan	Facilitative setting, organizational responsibilities, administrative tools, and connections to overarching national policies
Cultivate dedication toward implementation	Governmental endorsement, acknowledgment by stakeholders, and pinpointed financial resources
Implement frameworks	Framework for the development of water infrastructure and enhance capacity
Supervise and assess the advancement	Metrics tracking advancements in IWRM and benchmarks for water infrastructure development frameworks

14.7.3 CHALLENGES IN THE IMPLEMENTATION OF IWRM

IWRM planning and implementation challenges fall into four main categories, according to Roy et al. (2011). First, political difficulties result from the requirement to foster agreement and dedication among participants, particularly in transboundary basins. For IWRM to be successful, political will and collaboration are essential. In addition, due to resource constraints and competing goals, financial challenges, such as funding accessibility and allocation, can make IWRM and sustainable water management more challenging. Third, overcoming technical challenges, including a lack of data and integrating other sectors into water management, calls for innovation, technology, and capacity building. In addition, institutional constraints include coordination, governance, and stakeholder issues that can be resolved by bolstering institutional frameworks, expanding stakeholder participation, and improving governance procedures to support sustainable water management initiatives.

14.8 CONCLUSION

The framework of IWRM is essential for tackling the many problems that arise in managing water resources. IWRM continues to be a crucial instrument as global water scarcity and pollution problems persist, challenging us to give priority to cooperation, innovation, and responsible stewardship to maintain the sustainable management of our most valuable resource. IWRM is a comprehensive approach to managing water resources that aims at establishing a balance between fair access, sustainable water usage, and environmental preservation. IWRM facilitates equal access to clean water, reducing inequality and benefiting both current and future generations. IWRM stands as a pivotal blueprint for achieving water resource sustainability in aquaculture. By embracing a systemic approach, involving all stakeholders, recognizing the social dimension, and addressing knowledge gaps, IWRM provides a comprehensive framework. Through principles such as reliable funding, equity, recognizing water as an economic good, empowering women, and government support, IWRM ensures a balanced approach that harmonizes social, economic, and environmental values. The trans- and multidisciplinary nature of IWRM further reinforces collaboration, making it an essential tool to navigate the complexities of water management in aquaculture, ultimately contributing to a more sustainable and resilient future for this vital sector.

KEYWORDS

- **economic efficiency**
- **integrated water resources management**
- **livelihoods**
- **sustainability**
- **aquaculture**

REFERENCES

Anukularmphai, A. (2010). Implementing integrated water resources management (IWRM): Based on Thailand's experience.

Biswas, A. K. (2004). Integrated water resources management: A reassessment: A water forum contribution. *Water International*, *29*(2), 248–256.

Gerlak, A. K., & Mukhtarov, F. (2015). 'Ways of knowing' water: integrated water resources management and water security as complementary discourses. *International Environmental Agreements: Politics, Law and Economics*, *15*, 257–272.

Global Water Partnership—Technical Advisory Committee (GWP-TAC), (2000). Integrated Water Resources Management. TAC Background Paper No. 4. GWP, Stockholm, Sweden.

Global Water Partnership. (2004). Catalyzing change: A handbook for developing integrated water resources management (IWRM) and water efficiency strategies. Global Water Partnership Secretariat.

Godinez-Madrigal, J., Van, Cauwenbergh, N., & van der Zaag, P. (2019). Production of competing water knowledge in the face of water crises: Revisiting the IWRM success story of the Lerma-Chapala basin, Mexico. *Geoforum, 103*, 3–15.

Hooper, B. (2005). *Integrated River Basin Governance*. IWA Publishing.

Kasbohm, J., Grothe, S., Steingrube, W., Lai, L., Ngân, L., Hồng, N. T., & Hương, N. Q. (2009). Integrated water resources management (IWRM)—An introduction. *Journal of Geology, Series B*, *33*, 3–14.

Marquez-Romance, A., Guevara-Perez, E., Mejia-Marcacuzco, A., (2022). Need for integrated water resources management. *Dyna Energía y Sostenibilidad*, *11*(1), 16. https://doi.org/10.6036/ES10470

Meran, G., Siehlow, M., von Hirschhausen, C., Meran, G., Siehlow, M., & von Hirschhausen, C. (2021). Integrated water resource management: Principles and applications. *The Economics of Water: Rules and Institutions*, 23–121. https://doi.org/10.1007/978-3-030-48485-9_3

Rahi, E., & Jazi, H. (2021). Integrated water resources management: A tool for sustainable development. *Future Engineering Journal*, *2*(1).

Roy, D., Barr, J., & Venema, H. D. (2011). Ecosystem approaches in integrated water resources management (IWRM): A review of transboundary river basins. United Nations Environment Programme and the International Institute for Sustainable Development.

Sun, F., Chen, M., & Chen, J. (2011). Integrated Management of Source Water Quantity and Quality for Human Health in a Changing World.

Xie, M. (2006, October). Integrated water resources management (IWRM)—Introduction to principles and practices. In *Africa Regional Workshop on IWRM,* Nairobi, October.

CHAPTER 15

Remote Sensing and Geographic Information Systems: Tool for Sustainable Aquaculture and Fisheries Development

SUMAN TAKAR[1*], UDAI RAM GURJAR[2], TARANG KUMAR SHAH[3], and MADHU SHARMA[3]

[1]*College of Fisheries, Maharana Pratap University of Agriculture and Technology, Udaipur, Rajasthan, India*

[2]*KrishiVigyan Kendra, Dr. Rajendra Prasad Central Agriculture University, East Champaran, Bihar, India*

[3]*Department of Fisheries, Dr. G. C. Negi College of Veterinary and Animal Sciences, Chaudhary Sarwan Kumar Himachal Pradesh Krishi Vishvavidyalaya, Palampur, Himachal Pradesh, India*

*Corresponding author

ABSTRACT

The sustainable management of fisheries and aquaculture is a crucial issue of global concern. Advanced in satellite remote sensing (RS) and geographic information systems (GISs) for the fisheries and aquaculture sectors have been developing, and operational use is required for their sustainable development and effective management. Globally, there is consensus to monitor ecosystem-based management to detect change in the aquatic ecosystem due to climate change or overfishing. Reliable and targeted advisories on fish aggregation zones will benefit fishers by reducing the time and effort required to locate fish schools. In addition, issues related

Advances in Freshwater Aquaculture: Technology for Sustainable Development. Tarang Kumar Shah, Madhu Sharma, Mohd. Ashraf Rather, & Avdhesh Kumar (Eds.)

DOI: 10.1201/9781779640857-15

to aquaculture management, including the various utilizations of estuarine waters, the influence of water quality on finfish and shellfish aquaculture, and the accessibility of habitats, can be analyzed using GISs. The Food and Agriculture Organization has launched GIS Fish, a web-based portal that can obtain global expertise in GISs, RS, and mapping as applied to the aquaculture and fisheries sector.

15.1 INTRODUCTION

Geographic information systems (GISs) were originally developed for land use applications but have since been adapted for marine environments (Goodchild, 2000). GIS is a vital tool for generating and utilizing spatial information, focusing on the locations and attributes of various features. It aids in the collection, analysis, and visualization of spatial data for a wide range of applications. By recording coordinates, GIS accurately identifies the positions of features, providing numerical values that describe their locations on Earth (Padua, 2018). In the past, the lack of spatial data on underwater organisms and habitats limited the practical use of GIS for marine analysis. However, when integrated with other analytical tools and models, GIS enhances spatial monitoring and research, leading to improved management practices (Datta, 2003). Advances in technology, such as hydroacoustic telemetry, side-scan sonar, and radio telemetry, have enabled biologists to track and monitor fish species and integrate these data into GIS programs to create detailed geographical representations. Similarly, a technology using satellite or aircraft-based sensor technologies, well-known as remote sensing (RS), is used to detect and classify objects on Earth and those on the surface, along with the oceans and atmosphere, based on electronic radiation signals. The passive and active instruments are used in RS (Earth observatory, 2018). Passive devices are used to detect natural energy released from observed sources, such as reflected sunlight. The passive RS instruments include radiometer, imaging radiometer, spectroradiometer, and spectrometer, whereas active devices are those used to illuminate the objects observed using electromagnetic radiation. The most common active remote sensors include radio detection and ranging (radar), light detection and ranging (LiDAR), scatterometer, and laser altimeter. RS has immense applications in measuring sea surface temperature, chlorophyll estimation, wind pattern studies, identification of upwelling, mixed layer depth estimation, and so on. RS practices are useful to manage fisheries in a sustainable manner while also directing fishing fleets to locate fish schools more efficiently (Klemas, 2013). The current technological advancement could be utilized to better plan and manage aquaculture and fisheries resources, for example, GIS and RS technologies. Therefore, the usage of these advanced technologies is worth exploring, especially in the procurement of the essential information for formulating suitable policies, approaches, and management strategies for medium and long-term sustainable exploitation of fisheries and aquaculture resources.

15.2 IMPORTANCE OF GIS IN FISHERIES

Fisheries planning and management encompass numerous spatial components and significant challenges/issues. Spatial components include resource movements and migrations, transportation networks, identification of fishing grounds, habitat loss, markets, and environmental degradation, all of which constitute critical spatial issues. Fisheries experts, managers of aquatic resources, and policymakers, especially in developing countries, face significant challenges. GISs can play a pivotal role in addressing these challenges by leveraging various spatial components. However, there is a general lack of awareness and understanding of GIS technology and its potential benefits for fisheries management. In the aquaculture sector, GISs have been utilized to evaluate the suitability of coastal regions for aquaculture purposes (Simms, 2002; Tarunamulia and Sammut, 2023). Additionally, GIS has been instrumental in managing issues such as the multiple usages of estuarine waters, the effect of water quality on shellfish leases, habitat availability, aquaculture, and the conflict between aquaculture activities and habitats of marine waterfowl (Simms, 2002). GIS has also been employed to evaluate intertidal areas, such as mudflats, for the development of brackish water aquaculture, particularly shrimp farming. For instance, GIS was employed to evaluate intertidal areas (mudflats) for brackish water aquaculture development, involving a comprehensive analysis based on crucial parameters like engineering, water and soil quality, infrastructure, social constraints, and meteorological conditions in India (Karthik et al., 2005).

15.3 RS AND GIS AS EFFECTIVE MANAGEMENT TOOLS

The monitoring and evaluation of spatially distributed factors such as resource abundance and composition, feeding and reproduction patterns, nurseries, fishing efforts, harvests, tagging and recaptures, recruitment, regulatory zoning, control and surveillance measures, conflicts between gear and fleets, and ecosystem conditions pose significant operational and management hurdles for fisheries. Satellite data are also effectively utilized in oceanographic and meteorological forecasting, enhancing scientific understanding and operational safety at sea (Santos, 2000). The primary advantage of GIS is its capacity to manipulate and overlay data in multiple ways, enabling numerous analytical functions that facilitate faster and more efficient decision-making in fisheries (Datta, 2013). The Food and Agriculture Organization has recently introduced GIS Fish, an online platform that offers comprehensive global insights into the application of GISs, RS, and mapping in the fields of aquaculture and inland fisheries.

15.4 ROLE OF RS AND GIS IN FISHERIES AND AQUACULTURE

15.4.1 IDENTIFICATION OF POTENTIAL FISHING ZONES (PFZ)

RS of ocean color is the sole technique capable of presenting a comprehensive view of the marine autotrophic community's abundance, which forms the foundation of the marine food web. High phytoplankton biomass and temperature fronts have been correlated with increased fish abundance (Wanchana and Sayan, 2018). In India, forecasting information about potential fishing zones (PFZ) is provided free of charge to the artisanal fisherfolk community three times a week to enhance fishing efficiency and reduce fuel consumption (Santos, 2000). The locations of PFZs are disseminated to the fishing community through various media, often in the local language, to ensure accessibility and usability. The benefits of utilizing PFZ information are clearly demonstrated in the results, showing a noticeable increase in fish catch and a decrease in search time. When the dissemination of PFZ information is interrupted, it is usually due to the identification of nursery areas and peak spawning times, highlighting the need for a responsible fisheries forecasting program. Utilizing RS data from multiple satellites, the ESSO-Indian National Centre for Ocean Information Services provides daily advisories to fishermen, specifically referencing 586 fish landing centers along the Indian coast. Satellite images, collective with supplementary data, can be examined to find ocean conditions suitable for fish aggregation. The open ocean is biologically relatively less productive as compared to the coastal upwelling areas and shallow waters of the continental shelves. On continental shelves, nutrients are brought from rivers and full mixing of the water. There is also consistent upward movement of bottom water in upwelling zones leads to high productivity, which frequently charges the photic zone with nutrients (Behrenfeld et al., 2005). For example, the assessed mean productivities of the open sea, coastal region, and upwelling regions are 50, 100, and 300 g/cm^2/year, respectively. Hence, some of the world's largest fisheries are located in the upwelling regions, such as the main ones situated along the west coasts of North and South America, off the coast of Somalia, and the west coast of Africa (Pinet, 2009) (Figure 15.1).

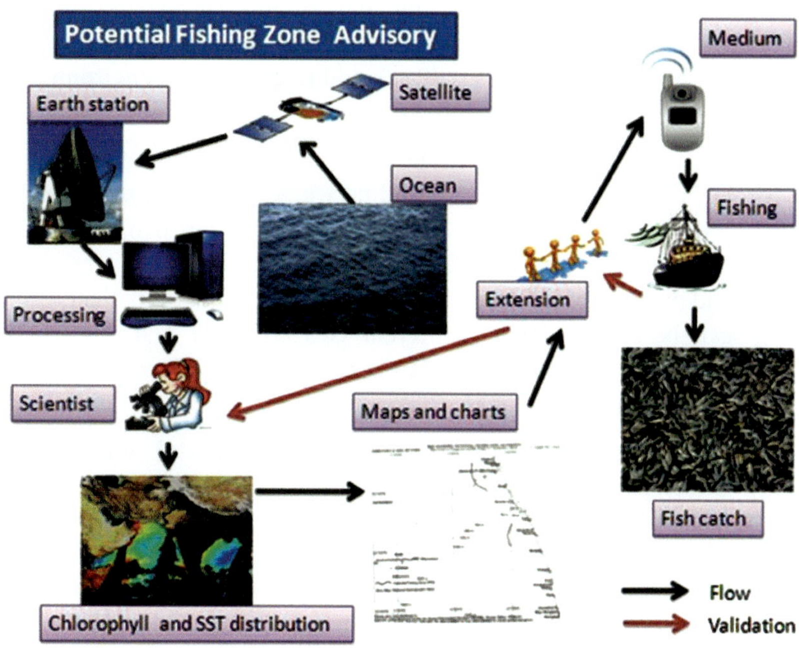

FIGURE 15.1 The process of potential fishing zone (PFZ) advisories as per ICAR-CCARI.

Source: ICAR-CCARI. https://ccari.res.in/Extension%20Folder%20No.%2070.pdf

15.4.2 SITE SELECTION FOR AQUACULTURE

This application of GIS in fisheries represents an innovative approach. Identifying a suitable site is crucial for successful aquaculture activities, as an optimal location facilitates better management of resources and ensures the sustainability of farming practices (Karthik et al., 2005). Several criteria, guidelines, and essential factors must be considered when selecting an aquaculture site. These include site topography, land slope, water flow and volume availability, water quality, weather conditions, access to utilities, and relevant legislation concerning water rights (Tarunamulia and Sammut, 2023). GIS provides an ideal platform to integrate this information, enabling the identification of areas that meet the optimal set of parameters for aquaculture (Padua, 2018).

15.5 TRACKING FISH ACTIVITY AND MOVEMENT

GIS is also employed for tracking and monitoring aquatic animals, analyzing their feeding and migration patterns, assessing the effectiveness of marine protected areas (MPAs), and responding to environmental degradation (Kaymaz and Yabanlı, 2017).

Analysis of fisheries catches and effort: Fishery managers are keenly interested in understanding where fishing effort is concentrated; how much the distribution of fish catches fish, what is the relationship between fish catch rate and fishing effort, and so on, and much exciting work is now being produced. Statistical results can be acquired from different GIS software. Catches can be elucidated in relation to different environmental factors or fish life cycles (Behrenfeld et al., 2005).

Matching fish distribution to environmental parameters: Understanding the relationship between fish distributions and a range of environmental factors is essential for fisheries management and the scientific community (Behrenfeld et al., 2005). The key parameters commonly studied include water temperature (especially thermal fronts), upwelling indices, marine chlorophyll abundance, water depth, salinity, and sediment composition.

Establishing regional and national fisheries databases: While not a direct GIS implementation to fisheries management, it is evident that substantial data inputs are essential for the functionality of fisheries GISs. Consequently, in some major fishery regions, significant efforts have been dedicated to establishing databases, metadata sets, and regional data centers, such as those in eastern Canada and various World Data Centres (Pinet, 2009). Fisheries-related datasets are gradually becoming accessible over the internet.

Mapping habitats and detecting changes: There are many advantages associated with RS, such as its synoptic view, multitemporal coverage, multispectral data, and cost-effectiveness, which make it indispensable for habitat detection and mapping. It provides a practical approach to gathering data from diverse and often inaccessible ecosystems, enabling the study of complex geographic terrains. The integration of GIS and RS has proven highly effective in monitoring and mapping coastal and marine ecosystems (Stelzenmüller et al., 2017). Since the launch of the USA's Landsat satellite in 1972, satellite imagery has been available for regions worldwide. This wealth of information facilitates the mapping and monitoring of changes in coastal areas, including the destruction of mangroves, land reclamation, and the shrinkage of lakes, estuaries, and other water bodies directly linked to the fisheries sector. Satellite imagery allows for the monitoring of different landscape features, helps quantify the rates of change, and detects significant changes in land cover. The application of GIS and RS is prevalent in the mapping and monitoring of habitats, including seagrasses, seaweeds, and coral reefs (Kaymaz and Yabanlı, 2017).

Conservation and management of marine resources: GIS-based studies provide a clear understanding of the spatiotemporal distribution of fish in the selected study area, aiding in the identification of critical fishing areas related to fisheries and marine biodiversity (Dineshbabu et al., 2014). Analyzing information collected in layers helps identify spawning periods and areas, juvenile segregation and migration, *in situ* growth of different species, resident taxa of specific fishing grounds, multigroup assemblages, and the basis for their gatherings and dependencies. For the first time in India, GIS is being applied to marine fisheries along the Karnataka coast through a project based on a database created for the trawl fisheries of this region (Biswas et al., 2018). Other uses of GIS include defining fish habitats, organizing and managing living marine resources (i.e., the dynamics of marine objects), and tracking marine mammals to analyze their hunting and migration patterns. The assessment of MPAs can help in addressing environmental degradation concerns (NOAA, 2017). Recently, participatory GIS applications have been developed for fisheries management, integrating local knowledge with verified scientific data. In these applications, marking specific locations or features based on a base map of the defined fishing grounds is crucial, and the involvement of resource users is essential.

Mapping of wetlands: It is crucial to conduct a detailed inventory of wetlands and their surrounding catchments in order to effectively conserve and manage wetland resources. Digital maps serve as powerful tools to accomplish this task. These maps are indispensable for monitoring and quantifying changes over time, thereby aiding in decision-making. The map preparation process begins with a ground survey (Burrough et al., 2015). The topographical maps produced by the Survey of India are the earliest accurate maps of India, depicting various land use and cover classes, including wetlands. RS technology is widely accepted as an essential tool for decision-making, visualization, analysis, and characterization of land, water, and atmospheric components. Ideally, the wetland boundary should encompass all open water and aquatic vegetation, such as submerged, floating, and emergent weeds. Satellite imagery should clearly indicate the extent of the wetland. The occurrence of vegetation in wetlands is essential for understanding their trophic conditions. The presence of vegetation in wetlands provides critical information about their trophic conditions. With the use of optical RS data, it is possible to accurately distinguish and map different types of floating and emergent vegetation.

15.6 ADVANTAGE OF RS AND GIS

- RS data used to reduce the scouting time for fish schools or locating PFZ, thus reducing fishing time and consumption of fuel.
- RS could provide the best possible advice in making fishery management decisions and developing efficient harvesting strategies for fishery resources.

- It is the best technique for understanding the primary productivity of the ocean at a given or specific time. This data can be used for the assessment of fish stocks.
- The satellite data could give information on weather, ensuring greater safety for the men out at sea.
- Reduce crew expenditures as a consequence of spending fewer days at sea during fishing.
- Reduce the costs of ship maintenance and improve safety at sea by using RS techniques.
- GIS offers a wide array of tools that ensure accurate outputs and comprehensive decision-making.
- Once maps are established in digital format, creating new maps, updating them, or merging them with other maps becomes a straightforward task.
- GIS meets the prerequisites for modeling scenarios in research and operational resource management tasks.
- It facilitates the seamless and immediate integration of other large datasets.
- GIS enables the display of spatially related data in an easily understandable and comprehensible manner.
- It allows for the production of high-quality, one-off maps, which would otherwise be cost-prohibitive.
- Multidisciplinary data utilized in GIS support a variety of studies and tasks, including global change analysis, satellite imagery rectification, change detection of marine processes, and visualization of these processes.
- GIS uses, particularly for the forecasting of large-scale oceanographic, dynamic atmospheric processes, and ecological processes.

15.7 CONCLUSIONS

For the sustainability of aquatic resources, proper management and monitoring of fish stocks and fish habitat are required. Therefore, satellite-derived information is being used to manage marine resources for more sustainable use and also to forecast the PFZ and monitor environmental parameters preferred by different species of fish (Subramanian et al., 2014). GIS tools can save billions of dollars annually through efficient route planning for the delivery of goods and services. By combining GIS and RS, we can assess fishery potential, evaluate the environmental health of river and lake basins, and allocate resources effectively between fisheries management and aquaculture development (Shih et al., 2017). Furthermore, GIS helps identify the sources, locations, and extents of environmental impacts and facilitates the development of practical plans for monitoring, managing, and mitigating such damage. While satellite data analysis assesses resource productivity, GIS contributes to creating comprehensive development plans for fisheries.

KEYWORDS

- **satellite**
- **ecosystem**
- **sustainable**
- **aquaculture**
- **fisheries**

REFERENCES

Behrenfeld, M. J., Boss, E, Siegel, D. A., & Shea, D. M. (2005). Carbon-based ocean productivity and phytoplankton physiology from space. *Global Biogeochemical Cycles*, *19*, GB1006. https://doi.org/10.1029/2004GB002299.

Biswas, S., Dineshbabu, A. P., Thomas, S., & Salian, T. S. (2018). *GIS and its Applications in Marine Fisheries Conservation and Management in Karnataka*, India. 2018.

Burrough, P. A., McDonnell, R., McDonnell, R. A. & Lloyd, C. D. (2015). *Principles of Geographical Information Systems*. Oxford University Press.

Datta, S. (2013). Application of GIS in Mapping of Fishery Resources. *Application of GIS in Mapping Fishery Resources*, 1–12.

Dineshbabu, A., Thomas, S., & Rohit, P. (2014). Gis-based spatial data analysis for marine fisheries management as a prerequisite for mariculture development. *Fishing Chimes*, *33*(10–1), 91–93.

Earth Observatory. (2018). https://earthobservatory.nasa.gov/Features/RemoteSensing/remote_0 8.php. 23 May, 2018.

Goodchild, M. F. (2000). GIS and transportation: Status and challenges. *GeoInformatica*, *4*(2), 127–139.

Karthik, M., Suri, J., & Saharan, N., & Biradar, R. S. (2005). Brackish water aquaculture site selection in Palghar Taluka, thane district of Maharashtra, India, using the techniques of remote sensing and geographical information system. *Aquacultural Engineering, 32*(2), 285–302.

Kaymaz, Ş. M. & Yabanlı, M. (2017). A review: Applications of geographic information systems (GIS) in marine areas. *Journal of Aquaculture Engineering and Fisheries Research, 3*(4), 188–198.

Klemas, V. (2013). Fisheries applications of remote sensing: An overview. *Fisheries Research, 148* , 124–136.

NOAA (2017). Fisheries GIS. http://www.nmfs.noaa.gov/gis/. 27 February, 2017.

Padua, S. (2018). *Gis applications in aquatic environment.* In: *e-Training Manual Recent Advances in Marine Fisheries and Taxonomic Research in India*. ICAR-Central Marine Fisheries Research Institute: Kochi, pp. 1–3.

Pinet, P. R. (2009). *Invitation to Oceanography*, 5th ed. Jones and Bartlett: Sudbury, MA.

Santos, A. M. P. (2000). Fisheries oceanography using satellite and airborne remote sensing methods: A review. *Fisheries Research, 49*, 1–20.

Shih, Y. C. (2017). Integrated GIS and AHP for marine aquaculture site selection in Penghu Cove in Taiwan. *Journal of Coastal Zone Management, 20*(1), 1–6.

Simms, A. (2002). GIS and aquaculture: assessment of soft-shell clam sites. *Journal of Coastal Conservation, 8*(1), 35–48.

Stelzenmüller, V., Gimpel, A., Gopnik, M., & Gee, K. (2017). Aquaculture site-selection and marine spatial planning: The roles of GIS-based tools and models. *In Aquaculture Perspective of Multi-Use Sites in the Open Ocean.* Springer: Cham. pp. 131–148.

Subramanian, S., Sreekanth, G. B., Manjulekshmi, N., Singh, N. P., Kolwalkar, J., Patil, T., & Fernandes, P. M. (2014). A manual on the use of potential fishing zone (PFZ) forecast. *Technical bulletin, 40*, 27.

Tarunamulia, & Sammut, J. (2023). Application of GIS and fuzzy sets to small-scale site suitability assessment for extensive brackish water aquaculture. *Annals of GIS, 29*(4), 585–601.

Wanchana, W., & Sayan, S. (2018). Application of GIS and remote sensing for advancing sustainable fisheries management in Southeast Asia. *Fish for the People, 16*(1), 21–28.

Index

D

E

O

P

Q

For Product Safety Concerns and Information please contact our EU representative GPSR@taylorandfrancis.com Taylor & Francis Verlag GmbH, Kaufingerstraße 24, 80331 München, Germany